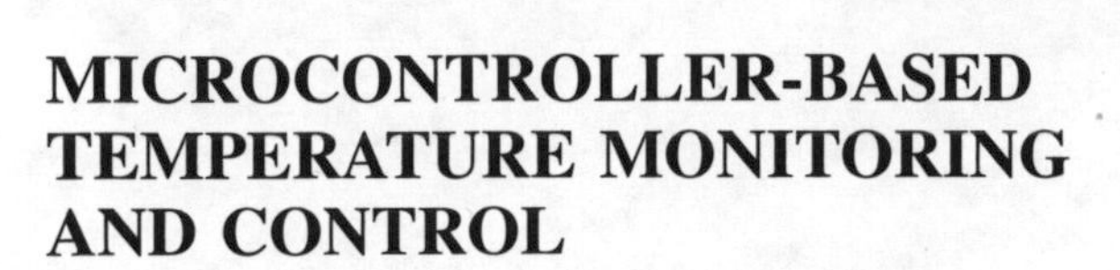

# MICROCONTROLLER-BASED TEMPERATURE MONITORING AND CONTROL

# MICROCONTROLLER-BASED TEMPERATURE MONITORING AND CONTROL

**Dogan Ibrahim**

Newnes

OXFORD AMSTERDAM BOSTON LONDON NEW YORK PARIS
SAN DIEGO SAN FRANCISCO SINGAPORE SYDNEY TOKYO

Newnes
An imprint of Elsevier Science
Linacre House, Jordan Hill, Oxford OX2 8DP
225 Wildwood Avenue, Woburn, MA 01801-2041

First published 2002



**British Library Cataloguing in Publication Data**
A catalogue record for this book is available from the British Library

**Library of Congress Cataloguing in Publication Data**
A catalogue record for this book is available from the Library of Congress

ISBN 0 7506 5556 9



Typeset by Laserwords Private Limited, Chennai, India
Printed and bound in Great Britain by Biddles Ltd, *www.biddles.co.uk*

# Contents

# Preface

Temperature measurement and control are vital in many industrial processes. Accurate control of the temperature is essential in nearly all chemical processes. In some applications, an accuracy of around 5–10°C may be acceptable. There are also some industrial applications which require better than ±1°C accuracy.

Temperature sensors come in many different forms and a number of techniques have evolved for the measurement of temperature. There are new forms of sensors which require no contact with the medium whose temperature is to be sensed. The majority of sensors still require to touch the solid, liquid, or the gas whose temperature is to be measured. Four technologies are currently in use: thermocouples (TCs), thermistors, resistance temperature detectors (RTDs), and IC sensors.

This book is an engineer's guide to planning, designing, and implementing temperature based control systems using a microcontroller. It will also prove invaluable for students and experimenters seeking real-world project work involving the use of a microcontroller. Engineers involved in the use of microcontrollers in measurement and control systems will find it an invaluable practical guide, providing design principles and application case studies backed up with sufficient control theory and electronics to develop their own systems. A basic mathematical and engineering background is assumed, but the use of microcontrollers is introduced from first principles. Exercises are provided at the end of most chapters to enable the reader to practice their knowledge.

This book is essentially in two parts: temperature measurement and temperature control. The early chapters are about the temperature sensors and their use to measure temperature. The basic theory behind the various temperature sensors and practical working microcontroller based temperature measuring and monitoring systems are described in detail. An introduction to the computer based temperature control systems and the Z-transformation is also covered and various digital control techniques are discussed with reference to the digital sampling theory.

Pre-requisites for the book are knowledge of mathematics, a programming language, and some classical control theory, which is usually possessed by most engineers during their undergraduate studies.

Chapter 1 is an introduction to the microcontrollers and their programming using a high-level language. The microcontrollers chosen in this book are the popular

PIC microcontroller series. The high-level language chosen for the examples is the widely used C language.

Chapter 2 is an introduction to the temperature scales and the concepts of temperature monitoring and control.

Chapter 3 describes the theory of thermocouple sensors. Types of thermocouples and their use in practical circuits are discussed. A microcontroller based thermocouple temperature measurement system is given with the full source code listing.

Chapter 4 describes the theory and practical applications of RTD temperature sensors. A microcontroller based RTD temperature measurement system is described with full circuit diagram and software listing.

Chapter 5 is about the thermistor temperature sensors and it describes the theory and gives practical circuits on the use of thermistors. In addition, a complete microcontroller temperature measurement system is described.

Chapter 6 describes the theory and applications of the analog and digital integrated circuit temperature sensors. Working practical projects are given for both types of sensors.

Chapter 7 is an introduction to the sampling process and digital control systems. The Z-transform is described and several examples are given. Block diagrams of sampled data systems are described with examples of block diagram manipulations. This chapter also describes the inverse Z-transform and the methods to get the time function of a sampled data system.

Chapter 8 is about the various computer based temperature control algorithms. Basic sampled data stability tests are described. In addition, the theory of the simple bang-bang type controller and the popular PID control algorithms are given together with the Ziegler–Nichols tuning procedures. This chapter also describes how the various control algorithms can be realized on a digital computer or a microcontroller.

Finally, Chapter 9 describes a complete microcontroller-based temperature control system. The system is modelled as a sampled data system and the block diagram of the modelled system is derived after performing simple unit tests. A microcontroller based PI and PID type digital controllers are then designed using the Ziegler–Nichols tuning algorithm. Full circuit diagram of the system and the software listings of the controllers are given.

# Chapter 1

# Microcomputer Systems

## 1.1 Introduction

The term microcomputer is used to describe a system that includes a minimum of a microprocessor, program memory, data memory, and input/output (I/O). Some microcomputer systems include additional components such as timers, counters, analogue-to-digital converters and so on. Thus, a microcomputer system can be anything from a large computer having hard disks, floppy disks and printers, to a single chip computer system.

In this book we are going to consider only the type of microcomputers that consist of a single silicon chip. Such microcomputer systems are also called microcontrollers.

## 1.2 Microcontroller systems

Microcontrollers are general purpose microprocessors which have additional parts that allow them to control external devices. Basically, a microcontroller executes a user program which is loaded in its program memory. Under the control of this program, data is received from external devices (inputs), manipulated and then data is sent to external output devices. A microcontroller is a very powerful tool that allows a designer to create sophisticated I/O data manipulation algorithms. Microcontrollers are classified by the number of bits in a data word. 8-bit microcontrollers are the most popular ones and are used in many applications. 16-bit and 32-bit microcontrollers are much more powerful, but usually more expensive and not required in many small to medium general purpose applications where microcontrollers are used.

The simplest microcontroller architecture consists of a microprocessor, memory, and I/O. The microprocessor consists of a central processing unit (CPU) and the control unit (CU).

The CPU is the brain of a microprocessor and is where all of the arithmetic and logical operations are performed. The control unit controls the internal operations of the microprocessor and sends out control signals to other parts of the microprocessor to carry out the required instructions.

Memory is an important part of a microcomputer system. Depending upon the application, we can classify memories into two groups: program memory and data memory. Program memory stores all the program code and this memory is usually non-volatile, i.e. data is not lost after the removal of power. Data memory is where the temporary user data is stored during the various arithmetic and logical operations. Data memories are usually volatile. There are basically five types of memories as summarized below.

### *RAM*

RAM means Random Access Memory. It is a general purpose memory which usually stores user data. RAM is volatile, i.e. data is lost after the removal of power. Most microcontrollers have some amount of internal RAM. Generally, 256 bytes is a common amount, although some microcontrollers have more, some less.

### *ROM*

ROM is Read Only Memory. This type of memory usually holds program or fixed user data. ROM memories are programmed at the factory and their contents cannot be changed by the user. ROM memories are only useful if you have developed a program and wish to order several thousands copies of it.

### *EPROM*

EPROM is Erasable Programmable Read Only Memory. This is similar to ROM but the EPROM can be programmed using a suitable programming device. EPROM memories have a small clear window on the chip where the data can be erased under an UV light. Many development versions of microcontrollers are manufactured with EPROM memories where the user program is usually stored. These memories are erased and re-programmed until the user is satisfied with the program. Some versions of EPROMs, known as OTP (One Time Programmable), can be programmed using a suitable programmer device but these memories cannot be erased. OTP memories cost much less than the EPROMs. OTP is useful after a project has been developed completely and it is required to make hundreds of copies of the program memory. Most microcontrollers have some amount of built-in EPROM. 2k bytes is a typical capacity, although some may have more and some less.

### *EEPROM*

EEPROM is Electrically Erasable Programmable Read Only Memory. These memories can be erased and also be programmed under program control. EEPROMs are used to save configuration information, maximum and minimum values, identification data etc. Some microcontrollers have built-in EEPROM memories (e.g. PIC16F84 contains a 64-byte EEPROM memory where each byte can be programmed and erased directly by software). EEPROM memories are usually very slow compared to other types of semiconductor memories.

### *Flash EEPROM*

This is another version of EEPROM type memory. This type of memory is becoming popular recently and is used in many microcontrollers (e.g. PIC16F84 contains 1k bytes of flash memory) to store the program data. The data on a Flash EEPROM is erased and then re-programmed using a programming device. The entire contents of the memory should be erased and then re-programmed. Flash EEPROMs are usually very fast.

One important distinction between a microcontroller and a microprocessor is that a microcontroller has special hardware in the form of input/output (I/O) ports for dealing with the outside world. I/O ports allow external signals and devices to be connected to the microcontroller. These ports are usually organized into groups of 8 bits and each group is given a name. For example, the PIC16F84 microcontroller contains two I/O ports named port A and port B. It is very common to have at least 8 I/O lines. Some microcontrollers have 32 or even 96 I/O lines, while some others may have only 6. On most microcontrollers the direction of the I/O port lines is programmable so that different bits can be programmed as inputs or outputs. Some microcontrollers provide bi-directional I/O ports where each port line can be used as either input or output. Some microcontrollers have "open-drain" outputs where the output transistors are left floating. External pull-up resistors are normally used with such output port lines.

## 1.3 Microcontroller features

Microcontrollers from different manufacturers have different architectures and different capabilities. Some may suit to a particular application while some others may be totally unsuitable. The hardware features of microcontrollers in general are described in this section.

### *Supply voltage*

Most microcontrollers operate with the standard +5 V supply. Some microcontrollers can operate at as low as +2.7 V and some will tolerate +6 V without any problems. You should check the manufacturers' data sheets about the allowed limits of the supply voltage.

### *The clock*

All microcontrollers require an oscillator (known as a clock) to operate. Most microcontrollers will operate with a crystal and two capacitors. Some will operate with resonators or with external resistor–capacitor pair. Some microcontrollers have built-in resistor–capacitor type oscillators and they do not require any external timing components (e.g. PIC12C672). Resonators are not as stable as the crystals but they are more stable than the resistor–capacitor networks. Crystal oscillators should be chosen for applications which require very accurate timing. For applications where the timing stability requirements are very modest, resonators should be chosen. If the application is not time sensitive you should consider using external or internal (if available) resistor–capacitor timing components for simplicity and low cost.

### *Timers*

Timers are an important part of any microcontroller. A timer is basically a counter which is driven from an accurate clock (or a division of this clock). Timers can be 8-bits or 16-bits long. Data can be loaded into the timers and they can be started and stopped under software control. Most timers can be configured to generate an interrupt when they reach a certain count (usually when they overflow). Some microcontrollers offer capture and compare facilities where a timer value can be read when an external event occurs, or the timer value can be compared to a preset value and interrupts can be generated when this value is reached. Timers can also be used to generate time delays in programs. It is typical to have at least one timer on every microcontroller. Some microcontrollers may have three or more while some others may have only two timers. The timer accuracy depends upon the type of clock device used and a crystal device should be chosen for very accurate timing applications.

### *Watchdog*

Many microcontrollers have at least one watchdog facility, also known as the *Watchdog Timer* (or WDT). A WDT is usually an 8-bit timer with a prescaler option and is clocked from a free running on-chip oscillator. The watchdog is usually refreshed by the user program at regular intervals and a reset occurs if the program fails to refresh the watchdog. Watchdog facilities are commonly used in real-time systems where it is required to check the proper termination of one or more activities. All PIC microcontrollers are equipped with a WDT.

### *Reset input*

This input resets the microcomputer. The reset logic is used to place the microcontroller into a known state. The source of the reset can usually be selected by the user and *Power-on Reset* (POR) is the most common form of reset in microcontrollers. Most microcontrollers have resistors connected to the supply voltage and this ensures that the microcontroller starts properly after the application of power (POR). Microcontroller manufacturers specify the state of the various registers after a reset signal is applied to a microcontroller. Some microcontrollers have internal reset circuitry which do not require any external components.

### *Interrupts*

Interrupts are a very important concept in microcontrollers. An interrupt causes a microcontroller to respond to external and internal (e.g. timer) events very quickly. When an interrupt occurs the microcontroller leaves its normal flow of execution and jumps directly to the Interrupt Service Routine (ISR). The source of an interrupt can either be internal or external. Internal interrupts are usually generated by the built-in timer circuits when the timer count reaches a certain value. External interrupts are generated by the devices connected external to the microcontroller and these interrupts are asynchronous, i.e. it is not known when an external interrupt will be generated. An example is the analogue-to-digital (A/D) conversion complete interrupt, which is generated when a conversion is completed. Interrupts can in general be nested such that a new interrupt can suspend the execution of another interrupt. Most microcontrollers have at least one, some have several interrupt sources. In some microcontrollers the interrupt sources can be prioritized so that a higher interrupt can suspend the execution of a lower interrupt service routine.

### *Brown-out detector*

Brown-out detectors are also common in many microcontrollers and they reset a microcontroller if the supply voltage falls below a nominal value. Brown-out detectors are usually employed to prevent unpredictable operation at low voltages, especially to protect the contents of EEPROM type memories. PIC microcontrollers are equipped with Brown-out detector circuits.

### *Analogue-to-digital converter*

Some microcontrollers are equipped with A/D converter circuits. Usually these converters are 8-bits, but some microcontrollers have 10- or even 12-bit converters. Some microcontrollers have multiple A/D channels (e.g. PIC16F877 is equipped with eight A/D channels). A/D converters usually generate interrupts when a conversion is complete so that the user program can read the converted data very quickly. A/D converters are very useful in control and monitoring applications since most sensors produce analogue output voltages.

### *Serial input/output*

Some microcontrollers contain hardware to implement serial asynchronous communications interface. The baud rate and the data format can usually be selected in software by the programmer. The built-in timer circuits are usually used to generate an accurate baud rate. If serial I/O hardware is not provided, it is easy to develop software to implement serial data transfer using any I/O pin of a microcontroller. Some microcontrollers incorporate SPI (Serial Peripheral Interface), $I^2C$ (Integrated InterConnect), or CAN (Controller Area Network) bus interfaces. These enable a microcontroller to interface to other compatible devices easily, usually over a suitable bus structure.

### *In-circuit programming*

In microcontroller development cycle, a microcontroller is normally removed from its socket and then programmed using a programmer device. The programmed chip is then re-inserted into its socket, ready for testing. This is usually very tedious work, especially during the development of complex software projects. In-circuit programming enables a microcontroller to be programmed while the chip is in the applications circuit, i.e. there is no need to remove the chip for programming. This feature speeds up the program development cycle considerably.

### *EEPROM data memory*

EEPROM type memory is also very common in many microcontrollers. The programmer can store non-volatile data in such memory and can also change this data whenever required. For example, if the microcontroller is used to measure the temperature, the maximum and minimum values during a period can be stored in an EEPROM type memory. Some microcontroller types provide between 64 and 256 bytes of EEPROM data memories, while some others do not have any such memories.

### *PWM output*

Some microcontrollers provide *Pulse-width Modulated* (PWM) outputs which can be used in some electronic applications. One such application is to provide an effective analog output from a microcontroller by varying the duty cycle of the PWM output. It is possible to modify the period or the duty cycle of a PWM output by loading the appropriate registers.

### *Capture–compare capability*

In some microcontrollers, the value of a timer register can be captured dynamically and then compared against a preset value. When the timer value is equivalent to the compared value, an output can be forced high or low.

### *LCD drivers*

LCD drivers enable a microcontroller to be connected to an external LCD display directly. This feature makes the displaying of data on LCD displays a very simple task. LCD drivers are not very common since most of the functions provided by them can be implemented by user software.

### *Analogue comparator*

Analogue comparators enable analogue signals to be compared easily. These circuits are not very common and are only implemented in some microcontrollers.

### *Real-time clock*

Real-time clock is another feature which is implemented in some microcontrollers. These microcontrollers usually keep the date and time of day and they are intended for the consumer market and for some real-time tasks.

### *Sleep mode*

Some microcontrollers (e.g. PIC) offer sleep modes where executing this instruction puts the microcontroller into a mode where the internal oscillator is stopped and the power consumption is extremely low. The devices usually wake up from the sleep mode by external reset or by a watchdog time-out.

### *Power-on reset*

Some microcontrollers (e.g. PIC) provide an on chip power-on reset circuitry which keeps the microcontroller in reset state until all the internal clock and the circuitry are initialized properly.

### *Low power operation*

Low power operation is important in portable battery operated applications. Some microcontrollers (e.g. PIC) can operate with less than 2 mA with 5 V supply, and around 15 $\mu$A at 3 V supply. Some other microcontrollers may consume as much as 80 mA or more at 5 V supply. The battery life can be extended by choosing a low power microcontroller, by operating the microcontroller at a low clock rate and by using low power logic gates in the applications circuit.

### *Current sink/source capability*

This is important if the microcontroller is to be connected to an external device, which draws large current for its operation. Some microcontrollers can sink and source only a few mA of current and driver circuits are required if they have to be connected to devices with large current requirements. PIC microcontrollers can sink and source up to 25 mA of current from each I/O pin which is suitable for most small applications, e.g. they can be connected to LEDs without any driver circuits.

## 1.4 Microcontroller architectures

Basically, two types of architectures are used in microcontrollers: *Von Neumann* architecture and *Harvard* architecture. Von Neumann architecture is used by a very large percentage of microcontrollers and here all memory space is on the same bus and instruction and data are treated identically. In the Harvard architecture (used by the PIC microcontrollers), code and data storage are on separate buses and this allows code and data to be fetched simultaneously, resulting in a more efficient implementation.

***RISC and CISC***

RISC (Reduced Instruction Set Computer) and CISC (Complex Instruction Set Computer) refer to the instruction set of a microcontroller. In a RISC microcontroller, instruction words are more than 8-bits wide (usually 12-, 14-, or 16-bits) and the instructions occupy one word in the program memory. RISC processors (e.g. PIC) have no more than about 35 instructions, and they offer higher speeds. CISC microcontrollers have 8-bit wide instructions and they usually have over 200 instructions. Some instructions (e.g. branch) occupy more than one program memory location.

## 1.5 The PIC microcontroller family

The microcontroller examples in this book are based on the popular PIC family of microcontrollers. In this section we shall look at the specifications of various PIC family members and also see how a microcontroller based project can be developed using the C high-level programming language.

The PIC families of microcontrollers are developed by Microchip Technology Inc. Currently they are some of the most popular microcontrollers, selling over 120 million devices each year. PIC microcontrollers have simple architectures and there are many versions of them, some with only small enhancements and some offering more features. The devices operate on 8-bit wide data and are housed in DIL and SOIC type 8-pin to 64-pin packages.

Basically, all PIC microcontrollers offer the following features:

- RISC instruction set with around 35 instructions
- Digital I/O ports
- On-chip timer with 8-bit prescaler
- Power-on reset
- Watchdog timer
- Power saving SLEEP mode
- Direct, indirect, and relative addressing modes
- External clock interface

- RAM data memory
- EPROM (or OTP) program memory

Some devices offer the following additional features:

- Analogue input channels
- Analogue comparators
- Additional timer circuits
- EEPROM data memory
- Flash EEPROM program memory
- External and timer interrupts
- In-circuit programming
- Internal oscillator
- USART serial interface

There are basically four families of PIC microcontrollers:

PIC12CXXX 12/14-bit program word

PIC16C5X 12-bit program word

PIC16CXXX and PIC16FXXX 14-bit program word

PIC17CXXX and PIC18CXXX 16-bit program word

Table 1.1 to Table 1.4 give a summary of the features of popular PIC microcontrollers. Some selected devices are described here briefly from each family.

**Family PIC12CXXX**

**PIC12C508**: This is a low-cost, 8-pin device with 512 × 12 EPROM program memory and 25 bytes of RAM data memory. The device can operate at up to 4 MHz clock input and there are only 33 single word instructions. The device features a 6-pin I/O port, 8-bit timer, power-on reset, watchdog timer, and internal 4 MHz RC oscillator capability.

**Table 1.1** Some PIC12CXXX family members

| Microcontroller | Program memory | Data Ram | Max speed (MHz) | I/O ports | A/D converter |
|---|---|---|---|---|---|
| 12C508 | $512 \times 12$ | 25 | 4 | 6 | – |
| 12C672 | $2048 \times 14$ | 128 | 10 | 6 | 4 |
| 12CE518 | $512 \times 12$ | 25 | 4 | 6 | – |
| 12CE673 | $1024 \times 14$ | 128 | 10 | 6 | 4 |
| 12CE674 | $2048 \times 14$ | 128 | 10 | 6 | 4 |

**Table 1.2** Some PIC16C5X family members

| Microcontroller | Program memory | Data RAM | Max speed (MHz) | I/O ports | A/D converter |
|---|---|---|---|---|---|
| 16C54 | $512 \times 12$ | 25 | 20 | 12 | – |
| 16C55 | $512 \times 12$ | 24 | 20 | 20 | – |
| 16C57 | $2048 \times 12$ | 72 | 20 | 20 | – |
| 16C58A | $2048 \times 12$ | 73 | 20 | 12 | – |
| 16C505 | $1024 \times 12$ | 41 | 4 | 12 | – |

**Table 1.3** Some PIC16CXXX and PIC16FXXX family members

| Microcontroller | Program memory | Data RAM | Max speed (MHz) | I/O ports | A/D converter |
|---|---|---|---|---|---|
| PIC17C42 | $2048 \times 16$ | 232 | 33 | 33 | – |
| 16C554 | $512 \times 14$ | 80 | 20 | 13 | – |
| 16C64 | $2048 \times 14$ | 128 | 20 | 33 | – |
| 16C71 | $1024 \times 14$ | 68 | 20 | 13 | 4 |
| 16F877 | $8192 \times 14$ | 368 | 20 | 33 | 8 |
| 16F84 | $1024 \times 14$ | 68 | 10 | 13 | – |

The "CE" version of the family (e.g. PIC12CE518) offers an additional 16-byte EEPROM data memory.

The high end of this family include devices with 14-bit instruction sets (e.g. PIC12C672) which also have more data RAM and EPROM program memories.

**Table 1.4** Some PIC17CXXX and PIC18CXXX family members

| Microcontroller | Program memory | Data RAM | Max speed (MHz) | I/O ports | A/D converter |
|---|---|---|---|---|---|
| 17C43 | 4096 × 16 | 454 | 33 | 33 | – |
| 17C752 | 8192 × 16 | 678 | 33 | 50 | 12 |
| 18C242 | 8192 × 16 | 512 | 40 | 23 | 5 |
| 18C252 | 16384 × 16 | 1536 | 40 | 23 | 5 |
| 18C452 | 16384 × 16 | 1536 | 40 | 34 | 8 |

### Family PIC16C5X

**PIC16C54**: This is one of the earliest PIC microcontrollers. The device is 18-pin with a 512 × 12 EPROM program memory, 25 byte of data RAM, 12 I/O port pins, a timer, and a watchdog timer. The device can operate at up to 20 MHz clock input.

Some other members of this family, e.g. PIC16C56 has the same structure but more program memory (1024 × 12). PIC16C58 has more program memory (2048 × 12) and also more data memory (73 bytes of RAM).

### Family PIC16CXXX

**PIC16C554**: This microcontroller has similar architecture to the PIC16C54 but the instructions are 14-bits wide. The program memory is EPROM with 512 × 14 and the data memory is 80 bytes of RAM. There are 13 I/O port pins, a timer, and a watchdog timer.

Some other members of this family, e.g. PIC16C71 incorporates four channels of A/D converter, 1024 × 14 EPROM program memory, 36 bytes of data RAM, timer, and watchdog timer. PIC16F877 is a sophisticated microcontroller which offers eight channels of A/D converters, 8192 × 14 program memory, 368 bytes of data memory, 33 I/O port pins, USART, $I^2C$ bus interface, SPI bus interface, 3 timers, and a watchdog timer. PIC16F84 is a very popular microcontroller, offering 1024 × 14 flash EEPROM program memory, 68 bytes of data RAM, 64 bytes of EEPROM data memory, 13 I/O port pins, timer, and a watchdog timer.

PIC16F84 and PIC16F877 will be used in the example projects given in this book.

### Family PIC17CXXX and PIC18CXXX

**PIC17C42**: This microcontroller has a 2048 × 16 program memory. The data memory is 232 bytes. In addition, there are 33 I/O port pins, USART, 4 timers, a watchdog timer, 2 data capture registers, and PWM outputs. PIC 17C44 is similar but offers more program memory.

PIC18CXXX members of this family include PIC18C242 type microcontroller with 8192 × 16 program memory, 512 bytes of data memory, 23 I/O port pins, 5 A/D channels (10-bits wide), USART, $I^2C$, and SPI bus interfaces, PWM outputs, 4 timers, watchdog timer, compare and capture registers, and multiply instructions.

All memory of the PIC microcontroller family is internal and it is usually not very easy to expand the memory externally. No special hardware or software features are provided for expanding either the program memory or the data memory. The program memory is usually sufficient for small dedicated projects. However the data memory is generally small and may not be enough for medium to large projects unless a bigger and more expensive member of the family is chosen. For some large projects even this may not be enough and the designer may have to choose a microcontroller from a different manufacturer with a larger data memory, or a microcontroller where the data memory can easily be expanded (e.g. the Intel 8051 series).

## 1.6 Minimum PIC configuration

The minimum PIC configuration depends upon the type of microcontroller used, but in general two external parts are needed: reset circuit and oscillator circuit.

Reset is normally achieved by connecting a 4.7k pull-up resistor from the MCLR input to the supply voltage. Sometimes the voltage rises too slowly and the simple reset function may not work. In this case, the circuit shown in Fig. 1.1 should be used.

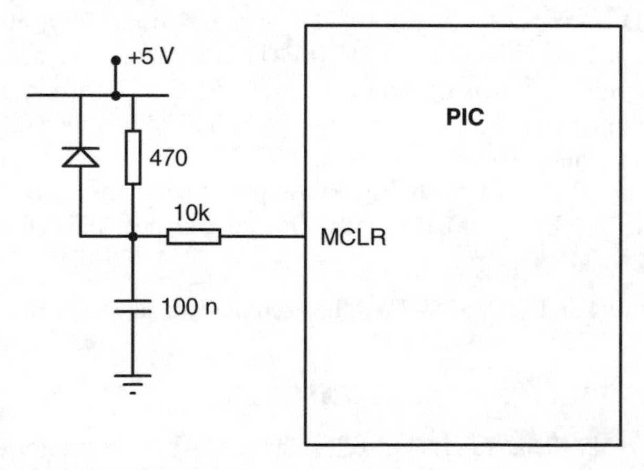


**Fig. 1.1** PIC reset circuit

PIC microcontrollers have built-in oscillator circuits and this oscillator can be operated in one of five modes:

- LP – Low power crystal
- XT – Crystal/resonator
- HS – High speed crystal/resonator
- RC – Resistor–capacitor
- No external components (only some PICs)

In LP, XT, or HS modes, an external oscillator can be connected to the OSC1 input as shown in Fig. 1.2.

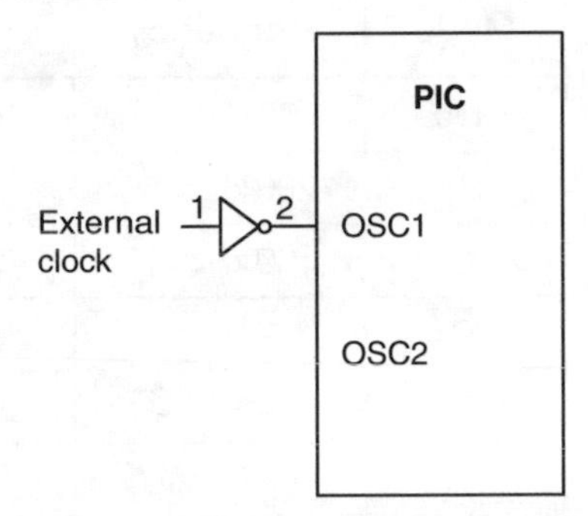


**Fig. 1.2** Using an external oscillator

The LP is the low power oscillator mode and is designed to trade speed for low power operation. The internal timing circuitry in the LP mode draws less current and this mode is optimum for low frequency operation.

The XT mode is designed to give a compromise between high frequency operation and modest power consumption. This mode is recommended for operation at up to 4 MHz.

The HS mode is also known as the high speed mode and this mode gives the highest frequency response. The current consumption is also higher than in the other modes. The HS mode is recommended for operation at up to 20 MHz.

### Crystal operation

As shown in Fig. 1.3, in this mode of operation an external crystal and two capacitors are connected to the OSC1 and OSC2 inputs of the microcontroller. The capacitors should be chosen as in Table 1.5. For example, with a crystal frequency of 4 MHz, two 22 pF capacitors can be used.

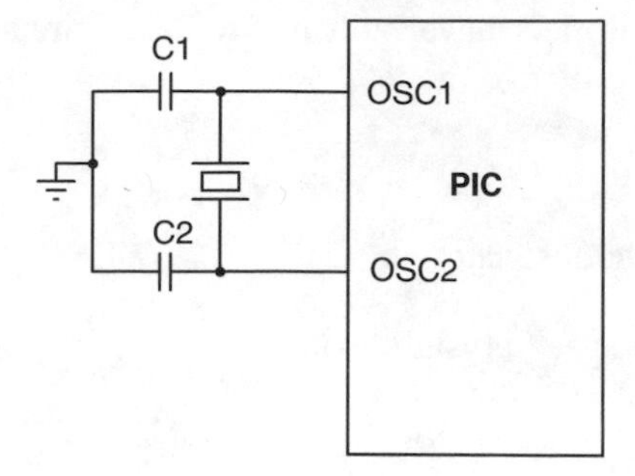


Fig. 1.3 Minimum PIC configuration with a crystal

Table 1.5 Capacitor selection for crystal operation

| Mode | Frequency | C1, C2 |
|---|---|---|
| LP | 32 kHz | 68–100 pF |
| LP | 200 kHz | 15–33 pF |
| XT | 100 kHz | 100–150 pF |
| XT | 2 MHz | 15–33 pF |
| XT | 4 MHz | 15–33 pF |
| HS | 4 MHz | 15–33 pF |
| HS | 10 MHz | 15–33 pF |

## Resonator operation

Resonators are available from 4 MHz to about 8 MHz. They are not as accurate as crystal based oscillators. Figure 1.4 shows how a resonator can be used with a PIC microcontroller.

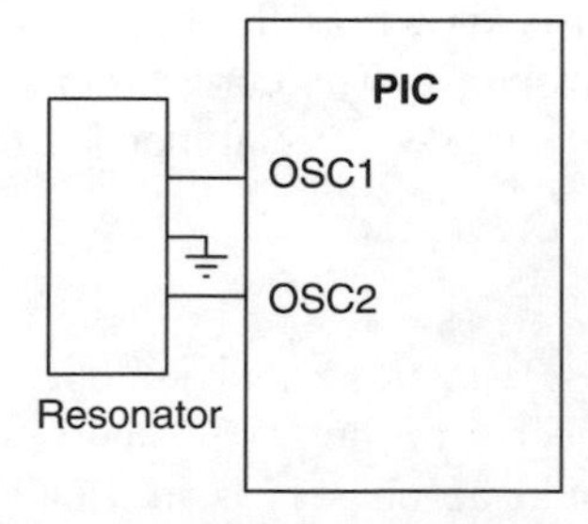


Fig. 1.4 Minimum PIC configuration with a resonator

### RC oscillator

For applications where the timing accuracy is not important we can connect an external resistor and a capacitor to the OSC1 input of the microcontroller as in Fig. 1.5. The oscillator frequency depends upon the values of the resistor and capacitor (see Table 1.6), the supply voltage, and the temperature. For most applications, using a 5k resistor with a 20 pF capacitor gives about 4 MHz and this may be acceptable.

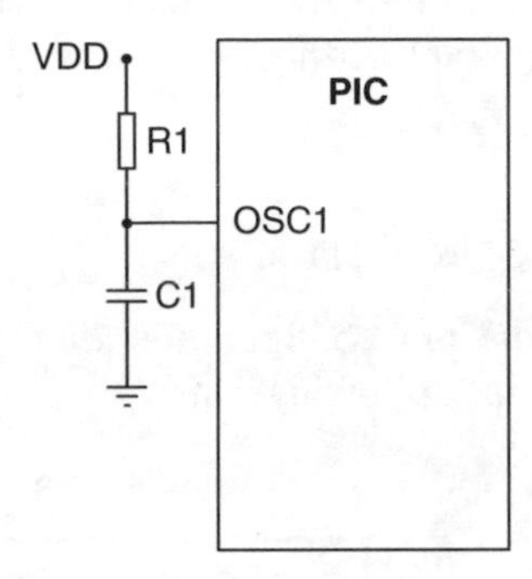


**Fig. 1.5** RC oscillator mode

**Table 1.6** RC oscillator component selection

| C1 | R1 | Frequency |
|---|---|---|
| 20 pF | 5k | 4.61 MHz |
| | 10k | 2.66 MHz |
| | 100k | 311 kHz |
| 100 pF | 5k | 1.34 MHz |
| | 10k | 756 kHz |
| | 100k | 82.8 kHz |
| 300 pF | 5k | 428 kHz |
| | 10k | 243 kHz |
| | 100k | 26.2 kHz |

### Internal oscillator

Some PIC microcontrollers (e.g. PIC12C672) have built-in complete oscillator circuits and they do not require any external timing components. The built-in oscillator is usually 4 MHz and is selected during programming of the device. The internal oscillator should be chosen for low cost applications.

## 1.7 PIC16F84 microcontroller

In this section we shall be looking at the architecture of the popular PIC16F84 microcontroller in greater detail. The architectures and instruction sets of other PIC microcontrollers are very similar and with the knowledge gained here we should be able to use any other PIC microcontroller without much difficulty.

Since we shall be programming in C language, there is no need to learn the exact details of the architecture or the instruction set. We shall only be looking at the features that a C programmer needs to know while developing software for such microcontrollers.

### 1.7.1 Pin configuration

Figure 1.6 shows the pin configuration of the PIC16F84. Basically this is an 18-pin device with the following pins:

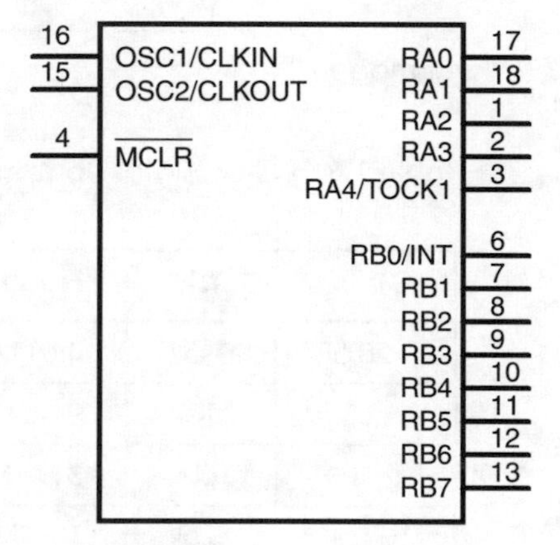


**Fig. 1.6** PIC16F84 pin configuration

RB0–RB7 Bi-directional port B pins

RA0–RA4 Bi-directional port A pins

VSS Ground

VDD Supply voltage

OSC1 Crystal or resonator (or external clock) input

OSC2 Crystal or resonator input

MCLR Reset input

INT This input is shared with RB0 and is also the external interrupt input

TOCK1 This input is shared with RA3 and is also the clock input for the timer

The microcontroller contains a 1024 × 14 Flash EEPROM, 68 bytes of data RAM, 64 bytes of EEPROM, 13 I/O port pins, timer, and a watchdog timer. Four interrupt sources are available:

- External INT pin interrupt
- Timer overflow interrupt
- Port B (4–7) interrupt on state change
- Data EEPROM write complete

PIC16F84 normally operates with a 5 V supply and consumes less than 2 mA current at 4 MHz. When operated at 2 V supply it consumes about 15 μA current at 32 kHz. The device is therefore extremely suitable for low-power portable applications.

*Register File Map* (RFM) is a layout of all the registers available in a microcontroller and this is extremely useful when programming the device, especially when using an assembler language. Figure 1.7 shows the RFM of the PIC16F84

| | BANK 0 | BANK 1 | |
|---|---|---|---|
| 00 | Indirect addr. | Indirect addr. | 80 |
| 01 | TMR0 | OPTION | 81 |
| 02 | PCL | PCL | 82 |
| 03 | STATUS | STATUS | 83 |
| 04 | FSR | FSR | 84 |
| 05 | PORTA | TRISA | 85 |
| 06 | PORTB | TRISB | 86 |
| 07 | – | – | 87 |
| 08 | EEDATA | EECON1 | 88 |
| 09 | EEADR | EECON2 | 89 |
| 0A | PCLATH | PCLATH | 8A |
| 0B | INTCON | INTCON | 8B |
| 0C | ***68 Bytes General Purpose Registers (0C to 4F)*** | ***Mapped (accesses in Bank 0)*** | 8C |
| ... | | | ... |
| ... | | | ... |
| ... | | | ... |
| 4F | | | CF |
| 50 | **50 to 7F not implemented** | | D0 |
| | | | |
| 7F | | | FF |

**Fig. 1.7** Register file map of PIC16F84

microcontroller. The RFM is divided into two parts: the *Special Function Registers* (SFRs), and the *General Purpose Registers* (GPRs). On the PIC16F84 device there are 68 GPRs and these are used to store temporary variables.

SFRs is a collection of registers used by the CPU and peripheral functions to control the internal device operations of the microcontroller. Depending upon the complexity of the devices, some other PIC microcontrollers may have more (or less) SFRs. It is important that the programmer understands the functions of the SFRs fully since they are used both in assembly language and in high-level languages.

The SFRs used while programming using a high-level language are described in the remainder of this section.

### 1.7.2 OPTION_REG register

The OPTION_REG register is a readable and writable register which contains various control bits to configure the on-chip timer and the watchdog timer. This register is at address 81 (hexadecimal) of the microcontroller and its bit definitions are given in Fig. 1.8. For example, to configure the INT pin so that external

| 7 | 6 | 5 | 4 | 3 | 2 | 1 | 0 |
|---|---|---|---|---|---|---|---|
| RBPU | INTEDG | TOCS | TOSE | PSA | PS2 | PS1 | PS0 |

Bit 7: PORTB Pull-up Enable
1 : PORTB pull-ups disabled
0 : PORTB pull-ups enabled

Bit 6: INT Interrupt Edge Detect
1 : Interrupt on rising edge of INT input
0 : Interrupt on falling edge of INT input

Bit 5: TMR0 clock source
1 : TOCK1 pulse
0 : Internal instruction cycle

Bit 4: TMR0 Source Edge Select
1 : Increment on HIGH to LOW of TOCK1
0 : Increment on LOW to HIGH on TOCK1

Bit 3: Prescaler Assignment
1 : Prescaler assigned to Watchdog timer
0 : Prescaler assigned to TMR0

Bit 2-0: Prescaler Rate
000 1:2
001 1:4
010 1:8
011 1:16
100 1:32
101 1:64
110 1:128
111 1:256

**Fig. 1.8** OPTION_REG bit definitions

interrupts are accepted on the rising edge of the INT pin, the following bit pattern should be loaded into the OPTION_REG:

X1XXXXXX

where X is a don't care bit and can be a 0 or a 1.

### 1.7.3 INTCON register

This register is readable and writable and contains the various bits for interrupt functions. This register is at address 0B and 8B (hexadecimal) of the microcontroller and the bit definitions are given in Fig. 1.9. For example, to enable interrupts so that external interrupt from pin INT can be accepted the following bit pattern should be loaded into register INTCON:

1XX1XXXX

| 7 | 6 | 5 | 4 | 3 | 2 | 1 | 0 |
|---|---|---|---|---|---|---|---|
| GIE | EEIE | TOIE | INTE | RBIE | TOIF | INTF | RBIF |

Bit 7: Global Interrupt Enable
1 : Enable all un-masked interrupts
0 : Disable all interrupts

Bit 6: EE Write Complete Interrupt
1 : Enable EE write complete interrupt
0 : Disable EE write complete interrupt

Bit 5: TMR0 Overflow Interrupt
1 : Enable TMR0 interrupt
0 : Disable TMR0 interrupt

Bit 4: INT External Interrupt
1 : Enable INT external interrupt
0 : Disable INT external interrupt

Bit 3: RB Port Change Interrupt
1 : Enable RB port change interrupt
0 : Disable RB port change interrupt

Bit 2: TMR0 Overflow Interrupt Flag
1 : TMR0 has overflowed (clear in software)
0 : TMR0 did not overflow

Bit 1: INT Interrupt Flag
1 : INT interrupt occurred
0 : INT interrupt did not occur

Bit 0: RB Port Change Interrupt Flag
1 : One or more of RB4-RB7 pins changed state
0 : None of RB4-RB7 changed state

**Fig. 1.9** INTCON bit definitions

### 1.7.4 TRISA and PORTA registers

Port A is a 5-bit wide port. Port pins 0 to 3 (i.e. RA0–RA3) have CMOS output drivers. Pin RA4 has an open drain output and should be connected to the supply voltage with a suitable pull-up resistor when used as an output pin. Each port pin has a direction control bit and this bit is stored in register TRISA. Setting a bit in TRISA makes the corresponding PORTA pin an input. Clearing a TRISA bit makes the corresponding PORTA pin an output. For example, to make bits 0 and 1 of port A input and the other bits output, we have to load the TRISA register with:

00000011

PORTA register address is 05 and TRISA address is 85 (hexadecimal).

### 1.7.5 TRISB and PORTB registers

Port B is an 8-bit wide bi-directional port. The corresponding data direction register is TRISB. A “0” in any TRISB position sets the corresponding port B pins to outputs. A “1” in any TRISB position configures the corresponding port B pins to be inputs. PORTB register address is at 06 and TRISB address is 86 (hexadecimal).

Some PIC microcontrollers have more than two ports and these additional ports are named as PORTC, PORTD etc. These ports also have direction registers named as TRISC, TRISD etc.

### 1.7.6 Timer module and TMR0 register

Timer is an 8-bit register (called TMR0) which can be used as a timer or as a counter. When used as a counter, the register increments each time a clock pulse is applied to pin TOCK1 of the microcontroller. When used as a timer, the register increments at a rate determined by the system clock frequency and a prescaler, selected by register OPTION_REG. Prescaler rates vary from 1:2 to 1:256. For example, when using a 4 MHz clock, the basic instruction cycle is 1 microsecond (the clock is internally divided by 4). If we select a prescaler rate of 1:16, the counter will be incremented at every 16 microseconds.

A timer interrupt is generated when the timer overflows from 255 to 0. This interrupt can be enabled or disabled by bit 5 of the INTCON register. Thus, if we require to generate interrupts at 200 microsecond intervals with a 4 MHz clock, we can select a prescaler value of 1:4 and enable timer interrupts. The timer clock rate is then 4 microseconds. For a timeout of 200 microseconds, we have to send 50 clocks to the timer. Thus, the TMR0 register should be loaded with $256 - 50 = 206$, i.e. a count of 50 before an overflow occurs.

TMR0 register has the address 01 which can be loaded from either an assembly language or from a high-level language.

PIC16F84 microcontroller contains 64 bytes of EEPROM data memory. This memory is controlled by registers EEDATA, EEADR, and EECON1. There are instructions in assembly language as well as in high-level languages to directly read and write data to this memory and thus these registers will not be discussed here.

PIC16F84 also contains a *Configuration Register* whose bits can be set or reset during the actual programming of the device. This register includes bits to enable or disable the following features:

- Enable/disable code protection
- Enable/disable power-on timer
- Enable/disable watchdog timer
- Source of the oscillator selection

Some other PIC microcontrollers may have A/D converters, PWM outputs, compare and capture registers and so on. The programming of these features are normally through the SFRs and you should find it easy once you have understood the principles described above.

# 1.8 PIC16F877 microcontroller

PIC16F877 is one of the microcontrollers that we shall be using in our projects. Some features of this microcontroller are described in this section.

## 1.8.1 Pin configuration

Figure 1.10 shows the pin configuration of the PIC16F877 microcontroller. Basically, this is a 40-pin device with the following pins:

| | |
|---|---|
| RB0/INT | Bi-directional port B pin. Also, external interrupt pin. |
| RB1–RB7 | Bi-directional port B pins. Some of these pins are also used as programming pins. |
| RA0–RA1 | Bi-directional port A pins. Also, analog input pins. |
| RA2 | Bi-directional port A pin. Also, analog input pin and analog negative reference voltage. |

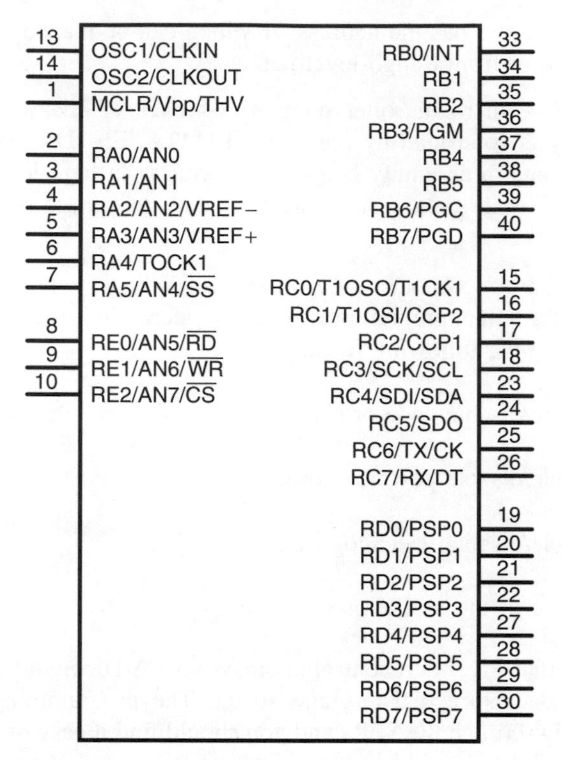


**Fig. 1.10** PIC16F877 pin configuration

| | |
|---|---|
| RA3 | Bi-directional port A pin. Also, analog input pin and analog positive reference voltage. |
| RA4 | Bi-directional port A pin. Also, clock input to Timer0 module. |
| RA5 | Bi-directional port A pin. Also, analog input and synchronous serial port slave select pin. |
| RE0–RE2 | Bi-directional port E pins. Also, analog inputs and parallel slave port control pins. |
| RD0–RD7 | Bi-directional port D pins. Also, parallel slave bus I/O pins. |
| VSS | Ground reference. |
| VDD | Positive supply voltage. |
| OSC1–OSC2 | Oscillator crystal inputs. |
| MCLR | Master clear input. |
| RC0 | Bi-directional port C pin. Also, Timer1 oscillator output or Timer1 clock input. |

| | |
|---|---|
| RC1 | Bi-directional port C pin. Also, Timer1 oscillator input or Capture2 Input/Compare2 output/PWM2 output. |
| RC2 | Bi-directional port C pin. Also, Capture 1 input/Compare1 output/PWM1 output. |
| RC3 | Bi-directional port C pin. Also, synchronous serial clock input/output. |
| RC4 | Bi-directional port C pin. Also the SPI Data In or $I^2C$ Data Out. |
| RC5 | Bi-directional port C pin. Also, SPI Data Out. |
| RC6 | Bi-directional port C pin. Also, the USART asynchronous transmit or synchronous clock. |
| RC7 | Bi-directional port C pin. Also, USART asynchronous receive or synchronous data. |

The PIC16F877 contains a $8192 \times 14$ Flash EEPROM, 368 bytes of data RAM, 256 bytes of EEPROM, 33 I/O port pins, 8 channels of A/D converters, $I^2C$ and SPI bus compatible pins, PWM output, capture and compare registers, parallel slave port pins, power-on reset, watchdog timer, power-saving sleep mode, brown-out detection circuitry, in-circuit programming support, USART, timers, and 14 sources of interrupts, including an external interrupt and various internal interrupt sources. The amount of program memory provided by the PIC16F877 should be sufficient for many temperature monitoring and control applications. These applications also need large data memories since most of the operations use non-integer, floating point arithmetic, requiring several bytes to store a single variable in the data memory.

Low level programming of the PIC16F877 is complex and the reader is referred to the manufacturer's data sheets. The device has a large number of registers with many options. Some important registers are described below. Further details will be provided when a particular register is used in an example project in the book.

### *OPTION_REG register*

This is a readable and writable register, which contains various control bits to configure the Timer0/watchdog prescaler, Timer0 clock source, port B pull-up enable bits, and external interrupt edge-trigger select bits.

### *INTCON register*

This is a readable and writable register, which contains various enable and flag bits for the Timer0 register, port B interrupt enable bits, peripheral interrupt enable bit, and the global interrupt enable bit.

***EECON1 register***

This is the EEPROM memory control register. The register contains the enable bits for EEPROM write, read, enable, and EEPROM error flag bits.

***T1CON register***

This is the Timer1 control register which contains the bit definitions for Timer1 clock source, oscillator enable control bit, and Timer1 prescaler select bits.

***ADCON0 register***

This register controls the operation of the A/D module. The bits in this register select the A/D clock source, one of eight analog channels, and the A/D status.

***ADCON1 register***

Register ADCON1 configures the functions of the A/D port pins and selects the A/D format.

## 1.9 Using C language to program PIC microcontrollers

Microcontrollers are usually programmed using the assembly language. This language consists of various mnemonics which describe the instructions. An assembler language is unique to a microcontroller and the assembly language of a certain microcontroller can not be used for any other type of microcontroller. Although the assembly language is very fast, it has some major disadvantages. Perhaps the most important disadvantage is that the assembly language can become very complex and difficult to maintain. It is usually a very time consuming task to develop large projects using the assembly language. Program debugging and testing are also considerably more complex, requiring more time.

Microcontrollers can be programmed using some of the popular high-level languages. For example, it is possible to use BASIC, PASCAL, or C compilers to program PIC family of microcontrollers. Most of these compilers generate native machine code which can directly be loaded into the memory of the target microcontroller.

In this section we shall be looking at the principles of programming PIC microcontrollers using the C high-level language. C is one of the most popular programming languages used today and there are several C compilers available for the PIC microcontrollers. In this section and in the example projects we shall be looking at how to program the PIC microcontrollers using one of the popular

C compilers known as the *FED C compiler*, developed by Forest Electronic Developments.

FED C is an efficient compiler which can be used to program all 16xxx and 12xxx series 14-bit core processors, including the popular PIC16F84 and the PIC16F877. The compiler is equipped with an integrated editor with syntax highlighting which makes the code development relatively easy. Additionally, an integrated simulator is provided which enables the programmer to test the code developed before loading it onto the microcontroller chip. A large number of library functions are provided which can easily be used by the programmer. Some of the reasons for choosing the FED C compiler are:

- Support for floating point arithmetic
- Direct support for LCD displays
- Availability of large number of mathematical functions
- Low cost

The program development cycle is relatively easy. User programs are normally developed on a PC using the integrated editor. The code can then be simulated using the integrated simulator. Finally, the correct code is loaded onto the required PIC microcontroller memory. Depending on the type of microcontroller used, either a flash memory programmer or an EPROM programmer device can be used to load the program memory of the target microcontroller.

### 1.9.1 FED C compiler variables

All the variables used in a program must be declared at the beginning of the program. FED C supports the following variable types:

*char*
This is used for 8-bit values which vary between −128 and +127.

*int*
int and short data types define variables which are 16-bits long, having values between the range −32768 to +32767.

*long*
A long variable is defined as being 32-bits. Therefore, it can store values from −2147483648 to +2147483647. Long arithmetic is usually slow and should be avoided if possible.

*unsigned*

The keyword *unsigned* can be used before any of the above variable types to make the variable unsigned. For example in the statement:

```
unsigned char k
```

Variable k is assumed to have values between 0 and 255.

*float*

Variable type float is used to define floating point variables. These variables are usually used in arithmetic and trigonometric calculations. The float type is a 32-bit. It consists of a sign bit, 8-bit exponent and 24-bit mantissa. The float number range is 1E −38 to 1E +38.

Table 1.7 is a summary of the FED C compiler variable types.

**Table 1.7** FED C compiler data types

| Variable type | No. of bytes used | Range |
|---|---|---|
| char | 1 | −128 to +127 |
| unsigned char | 1 | 0 to +255 |
| int | 2 | −32768 to +32767 |
| unsigned int | 2 | 0 to +65535 |
| short | 2 | −32767 to +32768 |
| unsigned short | 2 | 0 to +65535 |
| long | 4 | −2147483648 to +2147483647 |
| float | 4 | 1E −38 to 1E +38 |

All variables can be initialized when they are declared. In the following example, two character variables named w and x are declared and they are initialized to 5 and 9 respectively:

```
char w = 5, x = 9;
```

## 1.9.2 Comments in programs

Comments can be used anywhere in the programs. There are two ways of including comments in programs. Either by using a double forward slash character "//"

and writing the comment, or by enclosing the comment within the identifiers "/*" and "*/". An example is given below:

```
// This is a comment line

char x;          //This is also a comment line

/* This is a multi-line comment line
        The comment terminates on the next line
        Last line of comment */
```

### 1.9.3 Arrays

Arrays can be declared in FED C as in standard C. For example, the statement:

```
char dice[5];
```

Declares an array called *dice* with 5 variables of type character. The first variable is dice[0] and the last one is dice[4].

A string in C is a character array which is terminated with a null (ASCII 0) character. Strings can be initialized by enclosing the items in quotation marks. In the following example, string name is initialized to "Jones":

```
char name[] = "Jones";
```

Notice that the compiler can calculate the correct size of the array from the listed data items.

Arrays can have more than one dimension as shown in the following example:

```
char temp[3][2];
```

Here, *temp* is an array of 3 rows and 2 columns. The first item in the array is temp[0][0], the second item is temp[0][1], the third item is temp[1][0], the fourth item is temp[1][1], the fifth item is temp[2][0], and the last item is temp[2][1].

An array can be initialized by specifying the items in a curly bracket:

```
char p[3] = {3,6,9};
```

This sets p[0] to 3, p[1] to 6, and p[2] to 9.

Similarly, two dimensional arrays can be initialized as shown in the following example:

```
char q[2][3] = {{3,6,9},{2,4,6}};
```

### 1.9.4 Constants

Constants are a convenient way of declaring variables which are not expected to change within a program. A constant variable is declared by the keyword *const* as shown below:

```
const char clk;
const char cmd = 'X';
```

### 1.9.5 Enumerated constants

C enables assigning values to a group of constants. In the following example, three constants are declared: first takes the value 0, second takes the value 1, and last takes the value 2:

```
enum (first,second,last);
```

By default the first item is assigned the value 0 but this can be changed by equating any other value to it. For example:

```
enum (first=5,second,last);
```

In this example, first takes the value 5, second takes the value 6, and last takes the value 7.

### 1.9.6 Operators

The following operators can be used in FED C as in standard C:

| | |
|---|---|
| () [] | brackets |
| -> | structure member operations |
| ! | logical NOT |
| ~ | bit inversion |
| + – * / | arithmetic operators |
| % | modulus |
| ++ | increment |
| –– | decrement |
| & | address of |
| * | pointer |

| | |
|---|---|
| << >> | shift operators |
| >= > | comparison |
| <= < | comparison |
| sizeof | size of a variable |
| == != | logical equals and not equals |
| ‖ | bitwise OR |
| && | bitwise AND |
| ?: | conditional operator |
| += −= /= | assignment operators |
| \|= %= ^= | assignment operators |
| &= <<= >>= | assignment operators |

## 1.9.7 Program control in FED C

The following program control statements can be used in FED C:

if-else

for

while

do

goto

continue

break

switch and case

Control statements are important in every language and they are described here in more detail.

*If-else*
These statements are used for conditional operations. The general format of the if statement is:

```
if(condition) statement;
```

or,

```
if(condition)
{
        statement;
        statement;
        ................
}
```

The else statement is used to execute another set of statements when the condition is false. The general format is:

```
if(condition)
{
        statement;
        statement;
        .................
}
else
{
        statement;
        statement;
        .................
}
```

In the following example, if x is equal to 0 then b is incremented by 1, otherwise it is incremented by 2:

```
if(x = = 0)
        b = b + 1;
else
        b = b + 2;
```

*for*
The for control statement is used to perform iterations (or loops) in programs. The general format of this statement is:

```
for(start;condition;inc) statements;
```

or,

```
for(start;condition;inc)
{
        statement;
        statement;
        ................
}
```

where start is the starting value of the loop variable, condition is the condition that the loop variable is tested with, inc is the next value of the loop variable.

In the following example, variable cnt is incremented 10 times in a loop.

```
for(i=1;i<=10;i++) cnt++;
```

Here, the loop variable is i and it starts from 1. This variable is incremented at each iteration, until it reaches 10.

It is possible to nest the for loops as shown in the following example:

```
for(i=0;i<10;i++)
{
        for(j=0;j<5;j++)
        {
                k++;
                m = k;
        }
}
```

The inner loop is executed 5 times for each iteration of the outer loop. The outer loop is executed 10 times and is controlled with variable i.

*while*

The general format of the while control statement is:

```
while(condition) statement;
```

or,

```
while(condition)
{
        statement;
        statement;
        ..............
}
```

The statements are executed whilst the specified condition is true. If the condition is false at the beginning of the loop then the statements will never be executed. In the following example, the loop is executed 10 times:

```
i = 0;
while(i < 10)
{
        j++;
        i++;
}
```

*do*

This control statement is similar to the while statement, but the condition is tested at the end of the loop. As a result of this, the do loop executes at least once. The general format of this statement is:

```
do
{
        statement;
        statement;
        ...............
}
while(condition);
```

In the following example, the loop is executed 5 times:

```
j = 0;
do
```

```
{
        k++;
        j++;
}
while(j < 5);
```

*goto*

The goto statement causes an unconditional jump to a specified label in the program. A label can be an alphanumeric string terminated with a colon ":" character. Use of the goto statement should be avoided as it can result in unreadable and untidy code. An example use of the goto statement is given below:

```
back:
        if(i < 10)goto back;
```

*continue*

The continue statement is used to jump within a loop. If the loop is a while loop then it jumps back to the conditional to check whether or not continue running the loop. If the loop is a for loop then it jumps to the last part of the statement and then back to the conditional.

In the following code, the program jumps out of the loop when variable i is equal to 5:

```
i = 0;
flag = 1;
while(flag = 1)
{
        i++;
        if(i = = 5)
        {
                flag = 1;
                continue;
        }
        j = i;
}
```

*break*

This statement causes a jump out of the loop and the statement following the end of the loop is executed. An example is given below where the loop is terminated when variable j is equal to 5:

```
while(i< 10)
{
        j++;
        if(j = = 5) break;
}
```

*switch and case*

This statement is similar to multiple if–else statements. A value is taken and then compared to a number of options, performing a different operation for each option. The following code is an example of converting a 1-digit hexadecimal number between "A" and "F" into decimal. Assume that the number is in variable ch, and the result is stored in variable res:

```
switch (ch)
{
case 'A':    res = 65;
             break;
case 'B':    res = 66;
             break;
case 'C':    res = 67;
             break;
case 'D':    res = 68;
             break;
case 'E':    res = 69;
             break;
case 'F':    res = 70;
             break
default:     res = 0;
             break;
}
```

Notice that the default statement is executed if none of the conditions are matched (i.e. ch is not equal to "A" to "F").

### 1.9.8 Header files

A header file is available for every type of PIC microcontroller supported by the FED PIC compiler. The standard PIC file registers and bits within the file registers are included in these header files. The header files have the filenames "P16nnnn.h" where nnnn is the processor type. For example, the header file for the PIC16F84 microcontroller is included at the beginning of a program with the statement:

```
#include <P16F84.h>
```

### 1.9.9 PIC PORT commands

PIC microcontroller ports are bi-directional and a control register is used to tell whether each port pin is an input or an output. These registers are called the TRISx registers where x is the port identifier. For example, the control register for Port A is TRISA and for Port B it is TRISB. When a bit in the control register is set to 0 then the corresponding port pin is configured as an output, when it is set to 1, then the pin is an input.

The statement TRISB = 255 will set all pins of Port B as inputs. The statement k = PORTB will read the contents of Port B into variable called k.

The header file defines the ports as normal unsigned chars types and as structures. The structures are named as PA, PB, PC, and so on. For example, the following statement can be used to set bit 0 of Port B to 1:

```
PB.B0 = 1;
```

### 1.9.10 Built-in functions

The FED C compiler supports a large number of built-in functions. Some of the most commonly used functions are listed below:

| | |
|---|---|
| ClockDataIn | Used to clock in serial data at an input pin of the PIC microcontroller. |
| ClockDataOut | Used to clock out serial data at an output pin of the PIC microcontroller. |
| ReadEEData | Read the EEPROM memory data |
| WriteEEData | Write data to the EEPROM memory |

| | |
|---|---|
| LCD | Used to write data to an LCD display connected to the microcontroller |
| LCDString | Write the supplied string to the connected LCD display |
| cos, sin, tan | Trigonometric functions |
| log, log10 | Logarithm functions |
| pow | Returns x to the power of y · ($x^y$) |
| sqrt | Square root function |
| rand | Random number generator |
| SerialIn | Asynchronous serial data at an input of the PIC microcontroller |
| SerialOut | Asynchronous serial data output from an output port of the PIC microcontroller |

Additionally, there are various forms of *printf* functions to print numbers, characters, and strings in a format specified by the user. Also, a number of string functions are provided to copy strings, compare strings, return the length of strings and so on.

### 1.9.11 Using a LCD display

One thing all microcontrollers lack is some kind of video display. A video display would make a microcontroller much more user-friendly as it will enable text messages and numeric values to be output in a more versatile manner than the 7-segment displays, LEDs, or alphanumeric displays. Standard video displays require complex interfaces and their cost is relatively high. LCDs are alphanumeric (or graphical) displays which are frequently used in microcontroller based applications. These display devices come in different shapes and sizes. Some LCDs have 40 or more character lengths with the capability to display several lines. Some other LCD displays can be programmed to display graphic images. Some modules offer colour displays while some others incorporate back lighting so that they can be viewed in dimly lit conditions.

There are basically two types of LCDs as far as the interfacing technique is concerned: parallel LCDs and serial LCDs. Parallel LCDs (e.g. Hitachi HD44780 series) are connected to the microcontroller circuitry such that the data is transferred to the LCD unit using more than one data line, and four or eight data lines are very common. Serial LCDs are connected to a microcontroller using only one data line and data is usually transferred to the LCD using the standard RS-232 asynchronous data communication protocol. Serial LCDs are much easier to use but they usually cost more than the parallel ones. Parallel LCDs are used in the temperature projects in this book to show the value of the measured temperature.

The programming of a parallel LCD is usually a complex task and requires a good understanding of the internal operation of the LCDs, including the timing diagrams. Fortunately, FED C language provides special commands for displaying data on HD44780 type parallel LCDs. All the user has to do is connect the LCD to the appropriate I/O ports of the microcontroller and then use these special commands to simply send data to the LCD.

*HD44780 LCD module*

HD44780 is one of the most popular LCD modules used in industry and also by hobbyists. This module is monochrome and comes in different shapes and sizes. Modules with line lengths of 8, 16, 20, 24, 32, and 40 characters can be selected. Depending upon the model chosen, the display width can be selected as 1, 2, or 4 lines. The display provides a 14-pin connector to interface to the external world. Table 1.8 shows the pin configuration and pin functions for the LCD. Below is a summary of the pin functions.

**Table 1.8** Pin configuration of the HD44780 LCD module

| Pin no. | Name | Function |
|---|---|---|
| 1 | $V_{SS}$ | Ground |
| 2 | $V_{DD}$ | +ve supply |
| 3 | $V_{EE}$ | Contrast |
| 4 | RS | Register select |
| 5 | R/W | Read/write |
| 6 | E | Enable |
| 7 | D0 | Data bit 0 |
| 8 | D1 | Data bit 1 |
| 9 | D2 | Data bit 2 |
| 10 | D3 | Data bit 3 |
| 11 | D4 | Data bit 4 |
| 12 | D5 | Data bit 5 |
| 13 | D6 | Data bit 6 |
| 14 | D7 | Data bit 7 |

$V_{SS}$ is the 0 V supply or ground. The $V_{DD}$ pin should be connected to the positive supply. Although the manufacturers specify a 5 V d.c. supply, the modules will usually work with as low as 3 V or as high as 6 V.

Pin 3 is named as $V_{EE}$ and this is the contrast control pin. This pin is used to adjust the contrast of the device and it should be connected to a variable voltage supply. A potentiometer is usually connected between the power supply lines with its wiper arm connected to this pin so that the contrast can be adjusted. This pin can be connected to ground for most applications.

Pin 4 is the Register Select (RS) and when this pin is LOW, data transferred to the display is treated as commands. When RS is HIGH, character data can be transferred to and from the module.

Pin 5 is the Read/Write (R/W) line. This pin is pulled LOW in order to write commands or character data to the LCD module. When this pin is HIGH, character data or status information can be read from the module.

Pin 6 is the Enable (E) pin which is used to initiate the transfer of commands or data between the module and the microcontroller. When writing to the display, data is transferred only on the HIGH to LOW transition of this line. When reading from the display, data becomes available after the LOW to HIGH transition of the enable pin and this data remains valid as long as the enable pin is at logic HIGH.

Pins 7 to 14 are the eight data bus lines (D0 to D7). Data can be transferred between the microcontroller and the LCD unit using either a single 8-bit byte, or as two 4-bit nibbles. In the latter case, only the upper four data lines (D4 to D7) are used. The 4-bit mode has the advantage that fewer I/O lines are required to communicate with the LCD.

*Connecting the LCD to the microcontroller*

The LCD module is assumed by default to be connected to Port B of a PIC microcontroller. The pin connections are as follows:

| LCD module | Port B pins |
|---|---|
| RS | B1 |
| R/W | B2 |
| E | B3 |
| D4 | B4 |
| D5 | B5 |
| D6 | B6 |
| D7 | B7 |

The port connections can be changed by defining a constant integer LCDPORT. For example, the following statement specifies that the LCD is connected to Port C of the microcontroller:

```
const int LCDPORT = &PORTC;
```

The LCD data pins D4 to D7 are assumed to be connected to bits 4 to 7 of the selected port.

It is also possible to connect the E, RS, and R/W pins of the LCD to other pins of the microcontroller by defining integer constants. In the following example, it is assumed that the E pin is connected to bit-1 of Port D, RS pin to bit-2 of Port D, and the R/W pin to bit-3 of Port D:

```
const int LCDEPORT = &PORTD;
const int LCDEBIT = 1;
const int LCDRSPORT = &PORTD;
const int LCDRSBIT = 2;
const int LCDRWPORT = &PORTD;
const int LCDRWBIT = 3;
```

The functions used to send data and control to the LCD module are *LCD* and *LCDString*. Function LCD sends a control function to the module as given below:

| LCD statement | Function |
|---|---|
| LCD(-1) | Initialize the display to 1 line |
| LCD(-2) | Initialize the display to 2 lines |
| LCD(257) | Clear display and home the cursor |
| LCD(258) | Return the cursor to home position |
| LCD(256+128+N) | Return cursor to position N on line 1, where N = 0 is the first character position on line 1 |
| LCD(256+192+N) | Return cursor to position N on line 2, where N = 0 is the first character position on line 2 |

For example, the following statements initialize the LCD to 1 line of operation, clear the display, home the cursor, and displays the message "COMPUTER":

```
LCD(-1);                      // Initialize to 1 line
LCD(257);                     // Clear display and home cursor
LCDString("COMPUTER");        // Display message COMPUTER
```

### 1.9.12 Structures

A structure enables the programmer to group a number of related variables and form a new type. For example, the details of a person can normally be represented by the following C statements:

```
unsigned char name[10];
unsigned char surname[10];
unsigned int age;
```

The above statements can be combined into a structure as follows:

```
struct Personal
{
        unsigned char name[10];
        unsigned char surname[10];
        unsigned int age;
}
```

"Personal" now is a new data of type structure and it can be used to declare variables of this type. In the following example, variable "MyDetails" has the type "Personal":

```
Personal MyDetails;
```

To access the variables which are members of MyDetails, the dot operator (.) is used. For example, the age can be set to 20 as:

```
MyDetails.age = 20;
```

It is possible to have nested structures where structures can be declared within structures.

Arrays of structures can also be declared. For example, if we want to store the details of 50 people, we could declare:

```
Personal MyDetails[50];
```

The age of the 10th person can then be set to 15 as:

```
MyDetails[15].age = 15;
```

It is also possible using structures to assign a number of bits to variables. In the following examples, x and y are 1-bit wide, and z is 6-bits wide and the structure occupies only 1 byte in memory:

```
struct flags
{
        unsigned char x:1;
        unsigned char y:1;
        unsigned char z:6;
}
```

### 1.9.13 Unions

Unions are used to overlay structure definitions. All the variables in a union occupy the same memory area. An example union declaration is given below:

```
union flags
{
        char x;
        int y;
}
```

In this example, variables x and y occupy the same memory area and the size of this union is 2 bytes long, which is the size of the biggest member of the union. When variable y is loaded with a 2 byte value, variable x will have the same value as the low byte of y.

### 1.9.14 User functions

A function is an independent block of code which can be called by the main program. Functions are usually used when it is required to repeat an operation at various parts of the main code. A function may or may not return a value to the calling program. Functions are declared at the beginning of a program by terminating them with a set of brackets. In the following example, function mult returns an integer value and function LED does not return any values:

```
int mult();

void LED();
```

Functions are written by specifying the name of the function, followed by any parameters used by the function. In the following example, Port A is set to 1 and the function does not return any value:

```
void buzzer()
{
        PORTA = 1;
}
```

The following function returns the square of integer parameter y:

```
int square(int y)
{
        return y*y;
}
```

Functions can have local variables which are only valid inside the body of the function. In the following example, variable w is a local variable:

```
int square(int y)
{
        int w;
        w = y*y;
        return w;
}
```

If the same variable is declared both inside and outside a function, then it is the one inside the function which is used. Functions are called by writing the name of the function, followed by any parameters. In the following example, the main program calls function square to calculate the square of an integer number:

```
/* The function. Calculates the square of an integer */
int square(int a)
{
        return a*a;
}
```

```
/* The main program. Calculates the square of x and stores in y */
void main()
{
        int x,y;

        x = 4;
        y = square(x);
}
```

### 1.9.15 Pointers

A pointer holds the address of a variable. For example, if x is a variable, then a pointer to x holds the address of x. A pointer is declared by using a star (*) character in front of the name. In the following example, p is declared as a character pointer:

```
char *p;
```

Although p is declared to be a pointer, it does not currently point to anything. We can insert the symbol "&" in front of a variable to get the address of that variable. Thus, we can set our pointer p to point to the address of variable z as:

```
p = &z;
```

p now holds the address of variable z. We can access the contents of a variable whose address is known by using the "*" operator. In the following example, the value of variable z is set to 5:

```
*p = 5;
```

Pointers can be set to point to arrays and then access the elements of an array. The name of an array holds the address of the array and hence a pointer can be equated to the name as shown in the following example:

```
char x[10];
char *p;

p = x;          //point to array
*p = 0;         //first element = 0
p++;            //increment pointer
*p = 1;         //second element = 1
```

The example sets the first element of the array to 0 and the second element to 1.

### 1.9.16 The pre-processor

The pre-processor is part of the C compiler which runs before the main compiler and controls the compiler. Pre-processor commands begin with the "#" character. Some commonly used pre-processor commands are given in this section.

*Macros*

The "#define" keyword is used to define a macro where a symbol is replaced with another symbol or constant where ever it appears in a program. For example the following statements equate symbol delay to 100 and symbol clock to 12:

```
#define delay 100
#define clock 12
```

When these symbols are used in a program, they are replaced with their values. For example, in the statement:

```
x = delay + 10
```

variable x is assigned value 110.

*Include*

This pre-processor command is used to include a file for compilation. In the following example, file "P16F877.h" is included at the beginning of the compilation:

```
#include <P16F877.h>
```

Similarly, the command:

```
#include "myfile.h"
```

includes file "myfile.h" at the beginning of the compilation. Note that when angle brackets are used the compiler searches for the specified file in the standard directories used for the compiler. When a pair of quote characters are used the compiler starts the search from the user's project directory and then if the file is not found there, then the standard compiler directory is searched.

*Conditional compilation*

Conditional compilation can be useful when it is required to compile sections of a code depending upon the values of defines or other parameters. For example, consider the following code:

```
#ifdef CLOCK
  c = 1000;
#else
  c = 2000;
#endif
```

If symbol CLOCK is defined, the value of variable c is set to 1000, otherwise, it is set to 2000. Symbol CLOCK can be defined at the beginning of the program using the statement:

```
#define CLOCK
```

## 1.10 PIC C project development tools

Development of a PIC C project requires several development tools. The tools required depend on the type of application and the type of microcontroller used. In general, the following tools are required:

- A suitable C language compiler. In this book we shall be developing projects using the FED PIC C compiler.
- A suitable PIC microcontroller programmer device. There are many programmers available on the market for this purpose. The choice here depends upon the amount of money you wish to spend and also the types of microcontrollers you will be programming. For many low end microcontrollers you may like to choose the FED PIC programmer, manufactured by Forest Electronic Developments. This unit is powered from a mains adapter and is connected to a PC with a 9-way serial port cable. The unit has an on-board 40-pin socket and the following PIC microcontrollers can be programmed: 16C55X, 16C6X, 16C7X, 16C8X, 16F8X, 12C508, 12C509, 14000,12C67X, 16F87X, 16F62X, 16C77X, and 18CXXX. Other more expensive programmers may also be chosen which are capable of programming most of the PIC family.
- An EPROM eraser. This is only required if you will be using PIC devices with EPROM windows (e.g. 16C71, 12C672 etc.). For flash microcontrollers (e.g. 16F84, 16F877 etc.) there is no need to purchase an EPROM eraser.
- A PC for program development. This could either be a laptop PC or a standard desktop PC. The CPU speed or the system configuration are not important and most PCs can be used for this purpose.

- A bread-board or some other experimentation kit where you can build and test your projects.
- Depending upon your application, you will also require to purchase PIC microcontrollers, crystals, capacitors, resistors, LEDs, LCDs, and other components.

## 1.11 Structure of a microcontroller based C program

The structure of a C program developed for a microcontroller is basically the same as the structure of a standard C program, with a few minor changes. The structure of a typical microcontroller based C program is shown in Fig. 1.11. It is always advisable to describe the project at the beginning of a program using

```
/*****************************************************
*
*
*    PROJECT:        Give project name
*    FILE:           Give filename
*    DATE:           Date program was created
*    PROCESSOR:      Give target processor type
*    COMPILER:       Compiler used
*
*
* Describe here what the program does....
*****************************************************\

#include <P16F84.h>                  //include statements
#include........

int i,j,.......                      //global variables
char x,......

void func()                          //user functions
{
// comments...
     .......
     .......
}


void main()                          //main program
{
// comments...
     ........
     ........
     ........
     ........
     ........
}
```

**Fig. 1.11** Structure of a microcontroller C program

comment lines. The project name, filename, date, and the target processor type should also be included in this part of the program. Any microcontroller specific header files should then be included for the type of target processor used. The lines of the main program and any functions used should also contain comments to clarify the operation of the program.

**Example 1.1**

Write a C program to turn on and off an LED connected to bit 0 of Port B of a PIC16F84 type microcontroller. There should be a one second delay between each output.

**Solution 1.1**

The circuit diagram of this example is shown in Fig. 1.12. The clock is generated by using a 4 MHz crystal device. A small LED is connected to bit 0 of Port B through a 470 ohm current limiting resistor.

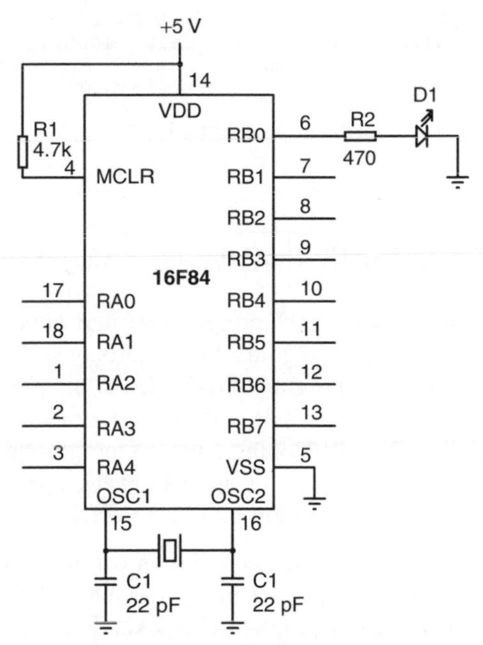


**Fig. 1.12** Circuit diagram of Example 1.1

The program listing is given in Fig. 1.13. Port B pins are configured as output with the statement TRISB = 0. An endless loop is formed with the *while* control statement and the LED is turned on and off inside this loop. The *wait* statement is used to generate a one second delay between each output.

```
/*******************************************************************
*
*
*       PROJECT:     PROJECT1
*       FILE:        PROJ1.C
*       DATE:        November 2001
*       PROCESSOR:   PIC16F84
*       COMPILER:    FED C
*
*
* This project turns ON and OFF an LED connected to bit 0 of Port B
* of the microcontroller.  A 1 second delay is inserted between each
* output.
*******************************************************************/

#include <P16F84.h>
#include <delays.h>

void main()
{
       TRISB=0;                    //Set all Port B pins as outputs
       while(1)                    //endless loop
       {
              PB.B0=0;             //set bit 0 of Port B OFF
              Wait(1000);          //delay 1000 ms
              PB.B0=1;             //set bit 0 of Port B ON
              Wait(1000);          //delay 1000 ms
       }
}
```

**Fig. 1.13** Program listing of Example 1.1

## 1.12 Program Description Language

There are many methods that a programmer may choose to describe the algorithm to be implemented by a program. Flowcharts have been used extensively in the past in many computer programming tasks. Although flowcharts are useful, they tend to create unstructured code and also a lot of time is usually wasted to draw them, especially when developing complex programmes. In this section we shall be looking at a different way of describing the operation of a program, namely by using a Program Description Language (PDL).

A PDL is an English-like language which can be used to describe the operation of a program. Although there are many variants of PDL, we shall be using simple constructs of PDL in our programming exercises, as described below.

### 1.12.1 START-END

Every PDL program (or sub-program) should start with a START statement and terminate with an END statement. The keywords in a PDL code should be highlighted in bold to make the code more clear. It is also a good practice to indent program statements between the PDL keywords.

**Example**

*START*

...........

...........

*END*

## 1.12.2 Sequencing

For normal sequencing in a program, write the steps as short English text as if you are describing the program.

**Example**

Turn on the valve

Clear the buffer

Turn on the LED

## 1.12.3 IF-THEN-ELSE-ENDIF

Use IF, THEN, ELSE, and ENDIF statements to describe the conditional flow of control in your programs.

**Example**

**IF** switch = 1 **THEN**

Turn on buzzer

**ELSE**

Turn off buzzer

Turn off LED

**ENDIF**

In this example, if the switch is equal to 1 then the buzzer is turned on, otherwise the buzzer and the LED are both turned off.

## 1.12.4 DO-ENDDO

Use DO and ENDDO control statements to show iteration in your PDL code.

**Example**

```
Turn on LED
DO 5 times
        Set clock to 1
        Set clock to 0
ENDDO
```

Variations of DO-ENDDO construct is to use other keywords like DO-FOREVER, DO UNTIL etc. as shown in the following examples.

```
Turn off the buzzer
IF switch = 1  THEN
        DO UNTIL Port 1 = 2
                Turn on LED
                Read port B
        ENDDO
ENDIF
```

or,

```
DO FOREVER
        Read data from port B
        Display data
        Delay a second
ENDDO
```

## 1.12.5 REPEAT-UNTIL

This is another useful control construct which can be used in PDL codes. The statements inside the loop are executed at least once. An example is shown below where the program loops until a switch value is equal to 1.

```
REPEAT
        Turn on buzzer
        Read switch value
UNTIL switch = 1
```

### 1.12.6 SELECT

This construct enables us to select an item and do various operations depending upon the value of this item. An example is shown below where if the variable *tmp* is 5, LED is turned ON, if *tmp* is 10, LED is turned OFF, if *tmp* is greater than 10 then a buzzer is turned on:

**SELECT** tmp

= 5

Turn on LED

= 10

Turn off LED

>10

Turn on buzzer

**END SELECT**

The FED C compiler fortunately supports most of the structured programming constructs.

## 1.13 Example LCD project

LCD displays will be used in later chapters to display the value of the measured temperature. A simple LCD based project is given here in order to illustrate the LCD interfacing and programming techniques.

Assume that it is required to develop a PIC16F84 microcontroller based project to send text messages to a HD44780 type LCD display. Connect the LCD to Port B of the microcontroller and develop a C program to send the text message "PIC LCD" to line 1 of the LCD.

The circuit diagram of this project is shown in Fig. 1.14. The LCD is connected to Port B of the PIC microcontroller as described in Section 1.9.11. The microcontroller clock is derived from a 4 MHz crystal. A 20k potentiometer is used to adjust the contrast of the LCD.

The following PDL describes the functions of the software:

**BEGIN**

Initialize LCD

Clear LCD and home cursor

Send text "PIC LCD" to the LCD

**END**

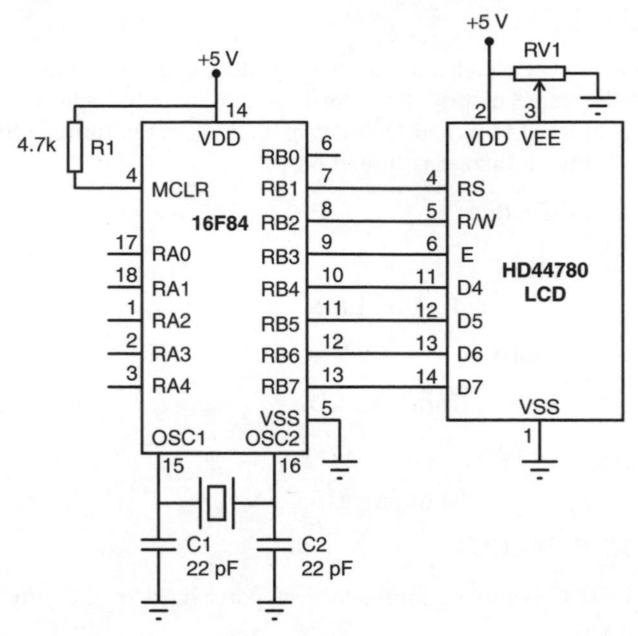


**Fig. 1.14** Circuit diagram of the LCD Project

```
/*****************************************************************
*
*
*     PROJECT:     PROJECT2
*     FILE:        PROJ2.C
*     DATE:        November 2001
*     PROCESSOR:   PIC16F84
*     COMPILER:    FED C
*
*
* This project sends the text message "PIC LCD" to line 1 of a
* LCD connected to PORT B of the PIC microcontroller.
*****************************************************************/

#include <P16F84.h>
#include <displays.h>

const int LCDPORT=&PORTB;             //Define LCD port as Port B

void main()
{
          LCD(-1);                    //Initialize LCD to 1 line
          LCD(257);                   //Clear display, home cursor
          LCDString("PIC LCD");       //Send text to LCD

          while(1)                    //wait here forever
          {
          }
}
```

**Fig. 1.15** Program listing of the LCD Project

The complete program listing is given in Fig. 1.15. Header file "*P16F84.h*" defines the PIC16F84 registers and port definitions. Header file "*displays.h*" defines the LCD configuration parameters. The program defines that the LCD is connected to Port B of the microcontroller. The LCD is then initialized, display cleared and the cursor set to the home position in line 1. The text message "PIC LCD" is sent to the LCD using statement *LCDString*. Finally, an endless loop is formed where the program waits forever.

## 1.14 Exercises

1. Explain the differences between a microprocessor and a microcontroller.

2. Explain the differences between an EPROM memory and a Flash EEPROM memory. Which one would you choose to store variable non-volatile data?

3. You are required to choose a clock source for a PIC microcontroller which is to be used for precision timing measurements and you have the options of using a crystal device or the internal RC oscillator. Which one would you choose? Why?

4. Calculate the required resistor and capacitor values to operate a PIC16F84 microcontroller from an internal 3 MHz clock.

5. Write a C program for a PIC16F84 microcontroller to read bit-2 of Port B and then send this data to an LED connected to bit-0 of the same port.

6. 8 LEDs are connected to port B output pins of a PIC16F84 microcontroller via suitable current limiting resistors. Write a program to count in binary from 0 to 255 and display the output on the LEDs. Insert a 500 ms delay between each output.

7. A HD44780 type single line parallel LCD is connected to a PIC16F84 microcontroller. Write a program to display all the ASCII characters on the LCD. Insert a 500 ms delay between each display.

8. A 7-segment LED display is to be controlled from a PIC16F84 microcontroller. Draw the circuit diagram to show how the LED can be connected to the microcontroller. Develop PDL statements to show how a number can be displayed on the 7-segment LED.

9. 8 LEDs are connected to Port B of a PIC16F84 microcontroller via suitable current limiting resistors and the LEDs are arranged in a circular manner. Develop PDL statements to show how the LEDs can be turned on and off in a clockwise fashion with only one LED on at any time.

10. A HD44780 type single line parallel LCD is connected to a PIC16F84 microcontroller. Write a program to count from 0 to 255 and display the output on the LCD display. Insert a one second delay between each output.
11. Develop PDL statements and a flow-chart for Exercise 10. Which method would you prefer? Why?

## 1.15 Further reading

The following books and reference manuals are useful in learning to program the PIC microcontrollers and the FED C compiler.

**PIC BASIC: Programming and Projects**, D. Ibrahim. Newnes, ISBN: 0 7506 5229 2

**Microcontroller Cookbook PIC and 8051**, M.R. James. Newnes, ISBN: 0 2405 1448 3

**PIC Microcontroller Project Book**, J. Iovine. McGraw-Hill, ISBN: 0 07 135479 4

**PIC Cookbook**, N. Gardner and P. Birnie. Character Press Ltd, ISBN: 1 899013 02 4

**50 Things To Do With a PIC**, Paul Benford. Bluebird Technical Press Ltd, ISBN: 1 901631 06 0

**Microchip Data on CDROM**, Microchip Technology Inc., 2355 W. Chandler Blvd. Chandler, AZ 85224. Website: *http://www.microchip.com*

**Learn to Use C With the Forest Electronic Developments PIC C Compiler**, Forest Electronics Developments, 60 Walkford Road, Christchurch, Dorset BH23 5QG

**Forest Electronic Developments PIC C Compiler**, (Address same as above)

# Chapter 2

# Temperature and its Measurement

Temperature is one of the most important parameters in process control. Accurate measurement of the temperature is not easy and to obtain accuracies better than 0.5°C great care is needed. Errors occur due to several sources, such as the sensor non-linearities, temperature gradients, calibration errors, and poor thermal contact.

This chapter describes the temperature scales and the types of sensors and their comparisons.

## 2.1 Temperature scales

The unit of the fundamental physical quantity known as the thermodynamic temperature (unit T) is the Kelvin, symbol K, defined as the fraction 1/273.15 of the thermodynamic temperature of the triple point of water. Most people think in terms of degrees Celsius. The relation between Kelvin and Celsius is:

$$T = {}^{\circ}C + 273.15 \tag{2-1}$$

From equation 2.1 it is clear that the triple point of water in degrees Celsius is 0.01°C. From a practical point of view, the ice point is 0°C, and the steam point 100°C. Table 2.1 gives the fixed temperature points of some commonly known physical phenomena.

**Table 2.1** Some commonly known temperature points

| Scale | Absolute zero | Ice point | Water boiling point |
|---|---|---|---|
| Celsius, °C | –273.15 | 0 | 100 |
| Fahrenheit, °F | –459.67 | 32 | 212 |
| Kelvin, °K | 0 | 273.15 | 373.15 |

On 1st January 1990, a new temperature scale was introduced, known as the *International Temperature Scale of 1990* (ITS-90). This scale supersedes the

International Practical Temperature Scale of 1968 (amended edition of 1975). The new ITS-90 scale resulted in changes of 1.5°C at 3000°C and less than 0.025°C between −100°C to +100°C. For the people who are interested in precision temperature measurement, these changes may be significant, but for most temperature applications, these changes are not important. Table 2.2 shows the fixed points adopted in ITS-90. The ITS-90 extends upwards from 0.65 K to the highest temperature practically measurable. It comprises a number of ranges and sub-ranges and several of these ranges or sub-ranges overlap. The melting and freezing point measurements are conducted at a pressure of 101.325 kPa.

**Table 2.2** Fixed points adopted in ITS-90

| Material | Measurement point | Temperature ($t_{90}$/K) | Temperature ($t_{90}$/°C) |
|---|---|---|---|
| Hydrogen | Triple point | 13.8033 | −259.3467 |
| Hydrogen | Boiling point at 33321.3 Pa | 17.035 | −256.115 |
| Hydrogen | Boiling point at 101292 Pa | 20.27 | −252.88 |
| Neon | Triple point | 24.5561 | −248.5939 |
| Oxygen | Triple point | 54.3584 | −218.7916 |
| Argon | Triple point | 83.8058 | −189.3442 |
| Mercury | Triple point | 234.3156 | −38.8344 |
| Water | Triple point | 273.16 | 0.01 |
| Gallium | Melting point | 302.9146 | 29.7646 |
| Indium | Freezing point | 429.7485 | 156.5985 |
| Tin | Freezing point | 505.078 | 231.928 |
| Zinc | Freezing point | 692.677 | 419.527 |
| Aluminium | Freezing point | 933.473 | 660.323 |
| Silver | Freezing point | 1234.93 | 961.78 |
| Gold | Freezing point | 1337.33 | 1064.18 |
| Copper | Freezing point | 1357.77 | 1084.62 |

## 2.2 Types of temperature sensors

There are many types of sensors to measure the temperature. Some sensors such as the thermocouples, RTDs, and thermistors are the older classical sensors and they are used extensively due to their big advantages. The new generation of sensors such as the integrated circuit sensors and radiation thermometry devices are popular only for limited applications.

The choice of a sensor depends on the accuracy, the temperature range, speed of response, thermal coupling, the environment (chemical, electrical, or physical), and the cost.

As shown in Table 2.3, thermocouples are best suited to very low and very high temperature measurements. The typical measuring range is −270°C to +2600°C. Thermocouples are low cost and very robust. They can be used in most chemical and physical environments. External power is not required to operate them and the typical accuracy is ±1°C.

**Table 2.3** Temperature sensors

| Sensor | Temperature range, °C | Accuracy ±°C | Cost | Robustness |
|---|---|---|---|---|
| Thermocouple | −270 to +2600 | 1 | Low | Very high |
| RTD | −200 to +600 | 0.2 | Medium | High |
| Thermistor | −50 to +200 | 0.2 | Low | Medium |
| Integrated circuit | −40 to +125 | 1 | Low | Low |

RTDs are used in medium range temperatures, ranging from −200°C to +600°C. They offer high accuracy, typically ±0.2°C. RTDs can usually be used in most chemical and physical environments, but they are not as robust as the thermocouples. The operation of RTDs require external power.

Thermistors are used in low to medium temperature applications, ranging from −50°C to +200°C. They are not as robust as the thermocouples or the RTDs and they can not easily be used in chemical environments. Thermistors are low cost and their accuracy is around ±0.2°C.

Semiconductor sensors are used in low temperature applications, ranging from −40°C to about +125°C. Their thermal coupling with the environment is not very good and the accuracy is around ±1°C. Semiconductors are low cost and some models offer digital outputs, enabling them to be directly connected to computer equipment without the need of A/D converters.

Radiation thermometry devices measure the radiation emitted by hot objects, based upon the emissivity of the object. But the emissivity is usually not known accurately, and additionally it may vary with time, making accurate conversion of radiation to temperature difficult. Also, radiation from outside the field of view may enter the measuring device, resulting in errors in the conversion. Radiation thermometry devices have the advantages that they can be used to measure temperatures in a wide range (450°C to 2000°C) with an accuracy better than 0.5%. Radiation thermometry requires special signal processing hardware and software and is not covered in this book.

The advantages and disadvantages of various types of temperature sensors are given in Table 2.4.

**Table 2.4** Comparison of temperature sensors

| Sensor | Advantages | Disadvantages |
|---|---|---|
| THERMOCOUPLE | Wide operating temperature range<br>Low cost<br>Rugged | Non-linear<br>Low sensitivity<br>Reference junction compensation required<br>Subject to electrical noise |
| RTD | Linear<br>Wide operating temperature range<br>High stability | Slow response time<br>Expensive<br>Current source required<br>Sensitive to shock |
| THERMISTOR | Fast response time<br>Low cost<br>Small size<br>Large change in resistance vs temperature | Non-linear<br>Current source required<br>Limited operating temperature range<br>Not easily interchangeable without re-calibration |
| INTEGRATED CIRCUIT | Highly linear<br>Low cost<br>Digital output sensors can be directly connected to a microprocessor without an A/D converter | Limited operating temperature range<br>Voltage or current source required<br>Self heating errors<br>Not good thermal coupling with the environment |

**Example 2.1**

It is required to measure the melting point of a chemical substance which is known to be between 1500°C and 1800°C. What type of temperature sensor would you choose?

**Solution 2.1**

The only sensor which can work at the required temperatures is a thermocouple. Care should be taken to ensure that the chosen thermocouple is suitable for the chemical environment.

**Example 2.2**

It is required to measure the temperature of a gas to an accuracy of around a few degrees. The temperature of the gas is between 0°C and +50°C. What type of temperature sensor would you choose?

**Solution 2.2**

Most sensors can be used for this purpose but probably the best choice would be to use a low cost semiconductor type sensor.

## 2.3 Measurement errors

There could be several sources of errors during the measurement of temperature. Some important errors are described in this section.

### 2.3.1 Calibration errors

Calibration errors can occur as a result of offset and linearity errors. These errors can drift with ageing and temperature cycling. It is recommended by the manufacturers to calibrate the measuring equipment from time to time. Sensor interchangeability is also an important criterion. This refers to the maximum likely error to occur after replacing a sensor with another of the same type without re-calibrating. RTDs are considered to be the most accurate and stable sensors.

### 2.3.2 Sensor self heating

RTDs, thermistors, and semiconductor sensors require an external power supply so that a reading can be taken. This external power can cause the sensor to heat, causing an error in the reading. The effect of self heating depends on the size of the sensor and the amount of power dissipated by the sensor. Self heating can be avoided by using the lowest possible external power, or by calibrating the self heating into the measurement.

### 2.3.3 Electrical noise

Electrical noise can introduce errors into the measurement. Thermocouples produce extremely low voltages and as a result of this, noise can easily enter into the measurement. This noise can be minimized by using low-pass filters, avoiding ground loops, and keeping the sensors and the lead wires away from electrical machinery.

### 2.3.4 Mechanical stress

Some sensors such as the RTDs are sensitive to mechanical stress and can give wrong outputs when subjected to stress. Mechanical stress can be minimized by avoiding the deformation of the sensor, by not using adhesives to fix a sensor to a surface, and by using sensors such as thermocouples which are less sensitive to mechanical stress.

### 2.3.5 Thermal coupling

It is important that the sensor used makes a good thermal contact with the measuring surface. If the surface has a thermal gradient (e.g. as a result of poor thermal conductivity) then the placement of the sensor should be chosen with care. If the sensor is used in a liquid, the liquid should be stirred to cause a uniform heat distribution. Semiconductor sensors usually suffer from good thermal contact since they are not easily mountable to the surface whose temperature is to be measured.

### 2.3.6 Sensor time constant

This can be another source of error. Every type of sensor has a time constant such that it takes time for a sensor to respond to a change in the external temperature. The time constant is defined as the time it takes for the output to reach 63% of its final steady-state value. Errors due to the sensor time constant can be minimized by improving the thermal coupling, or by using a sensor with a small time constant.

### 2.3.7 Sensor leads

Sensor leads are usually copper and therefore they are excellent heat conductors. These wires can lead to errors in measurements if placed in an environment with a temperature different to the measured surface temperature. These errors can be minimized by using thin wires, or by taking care in placing the lead wires.

## 2.4 Selecting a temperature sensor

Selecting the appropriate sensor is not always easy. This depends on factors such as the temperature range, required accuracy, environment, speed of response, ease of use, cost, interchangeability and so on. Traditionally, thermocouples are used in high temperature chemical industries such as glass and plastic processes. Environmental applications, electronics hobby market, and automotive industries generally use thermistors or integrated circuit sensors. RTDs are commonly used in lower temperature, higher precision chemical industries.

Chapter 3

# Thermocouple Temperature Sensors

Thermocouples are simple temperature sensors consisting of two dissimilar metals joined together. In 1821 a German physicist named Thomas Seebeck discovered that thermoelectric voltage is produced and an electric current flows in a closed circuit of two dissimilar metals if the two junctions are held at different temperatures. As shown in Fig. 3.1, one of the junctions is designated the hot junction and the other junction is designated as the cold or reference junction. The current developed in the closed loop is proportional to the types of metals used and the difference in temperature between the hot and the cold junctions.

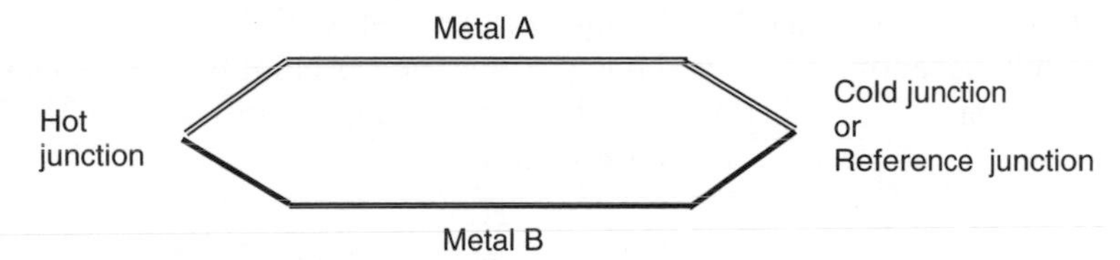


**Fig. 3.1** A thermocouple circuit

If the same temperature exists at both junctions, the voltages produced cancel each other out and no current flows in the circuit. A thermocouple therefore measures the temperature difference between the two junctions, and not the absolute temperature.

In order to measure the temperature we have to insert a voltage measuring device in the loop to measure the thermoelectric effect. Figure 3.2 shows such an arrangement where the measurement device is connected to the thermocouple with a pair of copper wires, using a terminal block.

Thermocouple wires are usually different metals from the measuring device wires and as a result, an additional pair of thermocouples are formed at the connection points. Figure 3.3 shows these additional undesirable thermocouples as junction 2 and junction 3. Although these additional thermocouples seem to cause a problem, the application of the Law of Intermediate Metals show that these thermocouples will have no effect if they are kept at the same temperature. The

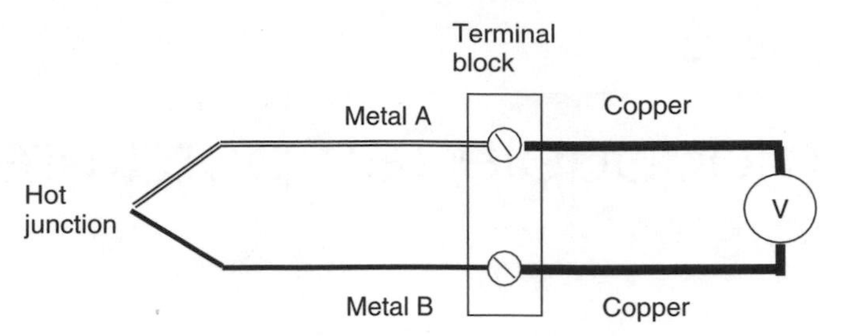


**Fig. 3.2** Connecting a measurement device

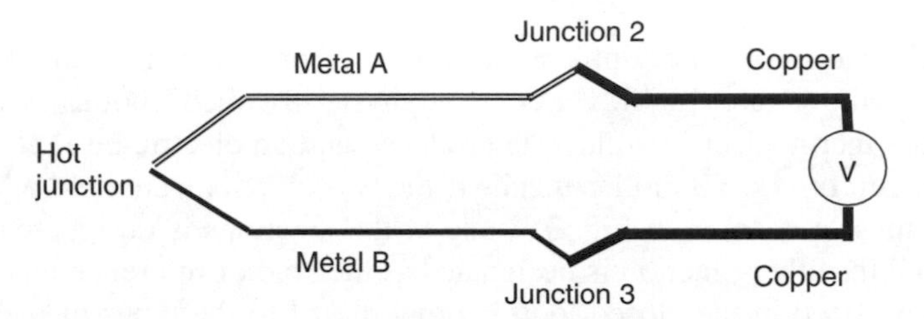


**Fig. 3.3** Additional thermocouples at junction 2 and junction 3

Law of Intermediate Metals simply states that a third metal may be inserted into a thermocouple system without affecting the system if the junctions with the third metal are kept isothermal (i.e. at the same temperature). Figure 3.4 illustrates the principle of the Law of Intermediate Metals.

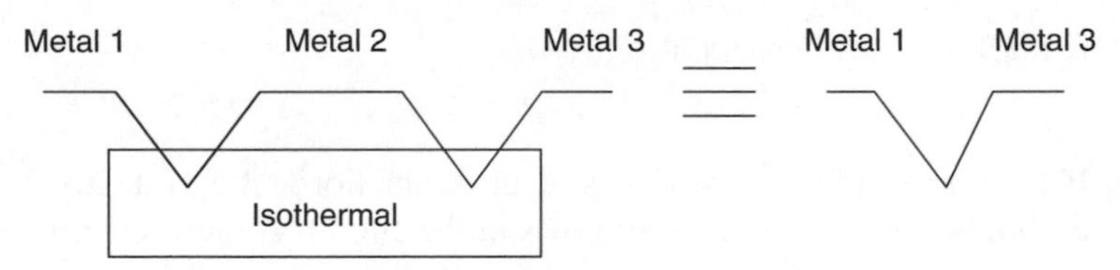


**Fig. 3.4** The Law of Intermediate Metals

Thus, if junction 2 and junction 3 are kept at the same temperatures, the voltage measured by the voltmeter will be proportional to the difference in temperature between junction 1 (hot junction) and junctions 2 and 3. This is illustrated in Fig. 3.5 where the junction temperature is at 150°C and the terminal block is kept at 50°C. The measured temperature is then the difference, i.e. 100°C. Junction 1 is the hot junction and the temperature of the terminal block is the temperature of the cold junction.

Thermocouples produce a voltage which is proportional to the difference in temperature between the hot and the cold (or the reference) junctions. If we want

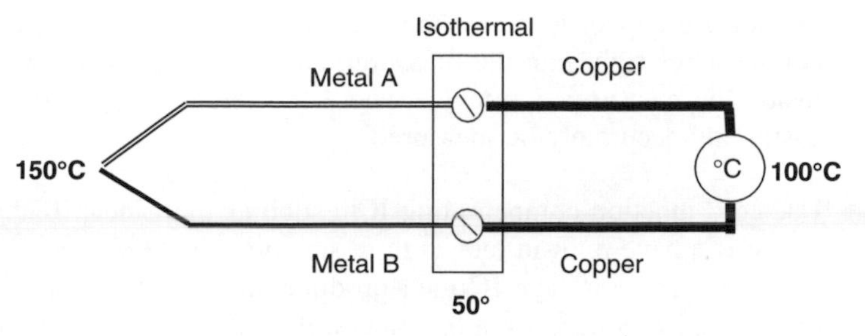


**Fig. 3.5** Thermocouple with isothermal terminal block

to know the absolute temperature of the hot junction, first we have to know the absolute temperature of the reference junction. If the reference junction is known and is controlled and stable then there is no problem. If the temperature of the reference junction is not known, one of the following methods can be used:

- Measure the temperature of the reference junction accurately and use this value to calculate the temperature of the hot junction. The simplest method to measure the temperature of the reference junction is to use a thermistor or a semiconductor temperature sensor. Then, the temperature of the reference junction should be added to the measured thermocouple temperature (see Fig. 3.5). This method gives accurate results and the cost is generally low.

- Locate the reference junction in a thermally controlled environment where the temperature is known accurately. For example, as shown in Fig. 3.6, an ice bath can be used to keep the reference junction at the ice temperature. Notice here that the reference junction is moved from the terminal block by inserting a Metal A into the measurement system. Alternatively, we could just immerse the terminal block into the ice, but this is not very practical. Ice bath compensation gives very accurate results but generally they are not very practical in industrial applications.

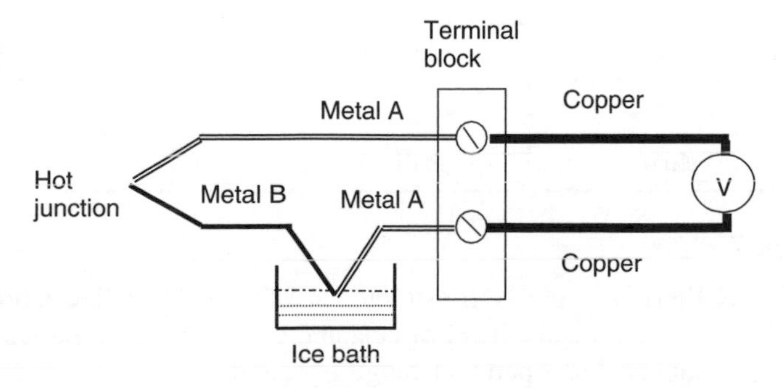


**Fig. 3.6** Using an ice bath for the reference junction

- Do not use copper wires for the measurement device, but extend the thermocouple wires right into the measurement device. Connect to the copper wires inside the measurement device where the reference junction temperature can easily and accurately be measured.

- Use cold junction compensating ICs, such as the Linear Technology LT1025. These ICs have built-in temperature sensors that detect the temperature of the reference junction. The IC then produces a voltage which is proportional to the voltage produced by a thermocouple with its hot junction at the ambient temperature and its cold junction at 0°C. This voltage is added to the voltage produced by the thermocouple and the net effect is as if the reference junction is kept at 0°C. Cold junction compensating ICs are accurate to a few degrees Celsius and they are used very commonly in many applications which do not require precision measurements.

## 3.1 Thermocouple types

There are about 12 standard thermocouple types which are commonly used. Each type is given an internationally approved letter that indicates the materials that the thermocouple is manufactured from. Table 3.1 shows the most popular thermocouples, their materials, and the usable temperature ranges.

**Table 3.1** Popular thermocouples

| Type | +Lead | −Lead | Seeback coefficient (μV/°C) | Temperature range (°C) |
|---|---|---|---|---|
| K | Ni + 10%Cr | Ni + 2%Al + 2%Mn + 1%Si | 42 | −180 to +1350 |
| J | Fe | Cu + 43%Ni | 54 | −180 to +750 |
| N | Ni + 14%Cr + 1.5%Si | Ni + 4.5%Si + 0.1%Mg | 30 | −270 to +1300 |
| T | Cu | Cu + 43%Ni | 46 | −250 to +400 |
| E | Ni + 10%Cr | Cu + 43%Ni | 68 | −40 to +900 |
| R | Pt + 13%Rh | Pt | 8 | −50 to +1700 |
| B | Pt + 30%Rh | Pt + 6%Rh | 1 | +100 to +1750 |

### *Type K*

Type K thermocouple is constructed using Ni-Cr (called Chromel), and Ni-Al (called Alumel) metals. It is low cost and one of the most popular general purpose thermocouples. The operating range is around −180°C to +1350°C. Sensitivity is approximately 42 μV/°C. Type K is most suited to oxidizing environments.

*Type J*

Constructed from iron and Cu-Ni metals, the temperature range of this thermocouple is −180°C to +750°C. As a result of the risk of iron oxidization, this thermocouple is used in plastic moulding industry. The sensitivity of type J thermocouple is 54 μV/°C. Type J thermocouple is generally recommended for new designs.

*Type N*

Type N thermocouple is constructed from Ni-Cr-Si (Nicrosil) and Ni-Si-Mg (Nisil) metals. The temperature range is −270°C to +1300°C. The sensitivity of type N is 30 μV/°C and it is generally used at high temperatures.

*Type T*

Type T thermocouple is constructed from Cu and Cu-Ni. The operating temperature range is −250°C to +400°C. This thermocouple is relatively low cost and is suited to low temperature applications. The sensitivity of this thermocouple is 46 μV/°C. Type T is tolerant to moisture.

*Type E*

Type E thermocouple is constructed using Ni-Cr (Chromel) and Cu-Ni (Constantan) metals. The temperature range is −40°C to +900°C. This thermocouple has the highest sensitivity at 68 μV/°C and it can be used in the vacuum and in unprotected sensor applications.

*Type R*

Type R is constructed using Pt-Rh (Platinum-radium) and Pt (Platinum). The sensitivity is low at 8 μV/°C. The temperature range of this thermocouple is −50°C to +1700°C. Type R is used to measure very high temperatures. Since it can easily be contaminated, it normally requires protection.

*Type B*

Type B thermocouple is constructed using Pt-Rh (Platinum-radium) metals with different compositions. The sensitivity is very low, at 1 μV/°C, and the temperature range is −100°C to +1750°C. Type B is used at measuring high temperatures, e.g. in the glass industry.

Thermocouples are usually identified by colour codes. Unfortunately, there is no standard single colour code and different countries have adopted different codes. For example, the American standard identifies the positive leg of a type K thermocouple with the yellow colour, and the negative leg with the red colour.

## 3.2 Thermocouple junction mounting

There are three common ways in which thermocouple junctions are mounted (see Fig. 3.7):

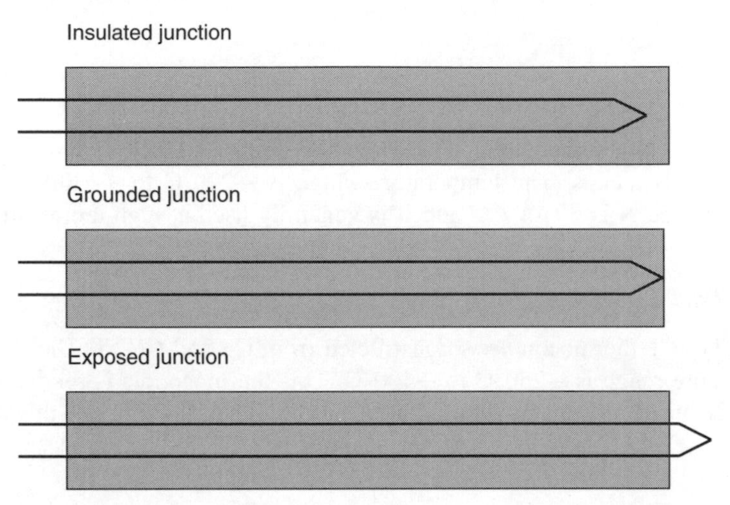


**Fig. 3.7** Thermocouple junction mounting

i. **Insulated junction**: Here, the junction is isolated from the sheath and as a result, the junction is suitable for corrosive environments and for high pressure applications. However, the response time is very long compared to other mounting techniques.

ii. **Grounded junction**: These junctions are also suitable for corrosive environments and high pressure applications. The response time is faster than the insulated junction, but the thermocouple is subject to ground loops and electrical noise.

iii. **Exposed junction**: Provides the fastest response. The junction is exposed and can easily be damaged. Not suitable for measurement in corrosive environments. This type of junction is generally used to measure the air and gas temperatures.

## 3.3 Thermocouple insulation

Although there are applications where the thermocouples can be used without any kind of insulation, in most industrial applications they must be protected from the environment or the media in which they are used to measure the temperature.

### 3.3.1 Standard insulating materials

There are no standards for the insulating materials for thermocouples, but the following insulators are commonly used:

**PVC**: can be used over the low temperature range, usually from −30°C to +100°C.

**Teflon**: offers a greater resistance to temperature and can be used over the temperature range of −250°C to +250°C.

**Glass fibre**: can be used over the temperature range of −50°C to +500°C.

### 3.3.2 Mineral insulated thermocouples

This is the most commonly used insulating and protection technique for thermocouples. The thermocouple wires are embedded in a densely packed insulant powder (typically magnesium oxide) which provides support and also insulates the thermocouple wires. Mineral insulated thermocouples have very good mechanical strength, long term stability, good insulation, fast response, and small size. The temperature range is very wide, extending from −200°C to +1250°.

## 3.4 Extension cables

It is sometimes required to use a thermocouple sensor at a remote location and the standard wire length supplied by the manufacturers may be too short. Extension cables are designed using the same conductors as the thermocouples so that problems do not occur at the connection points. These cables are not as expensive as the thermocouples themselves and they are designed to be flexible to make the extension easy. Extension cables are identified by inserting the letter "X" after the thermocouple type. For example, the extension cable for type K thermocouple is designated as "KX".

When very long extension cables are used, the cable resistance may become significant and tables are usually available from the manufacturers to calculate the lead resistance for a given thermocouple type and length.

## 3.5 Thermocouple response times

The thermocouple response time is defined as the time taken for the thermocouple to reach 63% of its final steady state value for a step change in temperature.

The response time depends upon several parameters, such as the thermocouple diameter, junction temperature, and construction of the junction. The response time of an exposed junction is the fastest. Also, the response time of an earthed junction is faster than that of an insulated junction. The response time becomes quicker when the diameter of the thermocouple is reduced. Also, a faster response time is obtained at lower junction temperatures. For example, the response time of a 3 mm diameter insulated junction is about 1.0 second when the junction temperature is 100°C. The response time of an earthed junction with the same diameter and temperature is 0.4 second. When the diameter of the thermocouple is reduced to 1.0 mm, the response time of an insulated junction becomes 0.16 second and that of an earthed junction becomes 0.07 second.

## 3.6 Thermocouple styles

The simplest thermocouple is a pair of bare wires with a welded junction and no harnessing and no special connectors. Such thermocouples are used with care and usually in laboratories. Industrial thermocouples are available in the following styles:

**Hand held probe**: These are general purpose thermocouple assemblies which can operate up to about 1100°C and are supplied with handles and a long extension cable with a miniature thermocouple plug at the end. Hand held probes are available in different tip configurations for measuring surface temperature, semi-solid materials (e.g. food, plastics etc.), penetration types (for pushing into materials such as frozen food), liquid measuring types, right-angled tips etc.

**Bolt thermocouple**: These thermocouples have screw heads which make them suitable to screw to the device whose temperature is to be measured, e.g. motors, industrial machines etc.

**Ring thermocouple**: These thermocouple assemblies are suitable for measuring the temperature of pipes, and other round shaped objects. The thermocouple junction is located within a ring.

**Washer thermocouple**: These thermocouple assemblies have a washer and they are suitable for measuring the temperature of vessels, pipes, and any kind of industrial equipment where the washer can be attached to.

**Thermocouple inserts**: Insert type sensors are suitable for insertion into a liquid (e.g. a tank) to measure the temperature. The sensor is housed in a stainless steel tube.

**Terminal heads**: There are many types of terminal heads available and they vary in size and shape. Terminal heads are widely used in process control applications.

## 3.7 Thermocouple temperature voltage relationships

There are several ways that we can describe the temperature voltage relationships of thermocouples and these are detailed in this section.

### 3.7.1 Using thermocouple reference tables

Thermocouple reference tables list the temperature and the generated thermocouple voltage for each thermocouple type when the reference junction is held at 0°C. The temperature is listed with an accuracy of 1°C and the thermocouple voltage is usually given in μV. These tables can be used as look-up tables in computer based thermocouple measurements.

Table 3.2 is an extract from the type K thermocouple reference table, covering the temperature range 0°C to +50°C.

**Table 3.2** Extract from type K thermocouple reference table

| °C | 0 | 1 | 2 | 3 | 4 | 5 | 6 | 7 | 8 | 9 |
|---|---|---|---|---|---|---|---|---|---|---|
| 0 | 0 | 39 | 79 | 119 | 158 | 198 | 238 | 277 | 317 | 357 |
| 10 | 397 | 437 | 477 | 517 | 557 | 597 | 637 | 677 | 718 | 758 |
| 20 | 798 | 838 | 879 | 919 | 960 | 1000 | 1041 | 1081 | 1122 | 1163 |
| 30 | 1203 | 1244 | 1285 | 1326 | 1366 | 1407 | 1448 | 1489 | 1530 | 1571 |
| 40 | 1612 | 1653 | 1694 | 1735 | 1776 | 1817 | 1858 | 1899 | 1941 | 1982 |
| 50 | 2023 | 2064 | 2106 | 2147 | 2188 | 2230 | 2271 | 2312 | 2354 | 2395 |

Voltages in μV

### 3.7.2 Using power series method

The temperature voltage relationship of thermocouples is non-linear and can be expressed as a polynomial (eqn. 3-1):

$$T = a_0 + a_1 v + a_2 v^2 + \cdots + a_n v^n \quad (3\text{-}1)$$

where T is the temperature (°C), $a_0, a_1, a_2, \ldots, a_n$ are coefficients which depend upon the type of thermocouple used, and v is the thermocouple voltage.

The coefficients $a_0, a_1, a_2, \ldots, a_n$ also depend on the temperature range and the coefficients of some popular thermocouples are given in Table 3.3 for the first 10 coefficients. The accuracy of the temperature calculation is improved when more coefficients are taken, but in general the first 10 coefficients are enough for an accuracy better than 1°C. In Table 3.3, the error given in the last row of

**Table 3.3** Thermocouple $a_i$ coefficients

| | **K**<br>**0°C to 500°** | **J**<br>**0°C to 760°C** | **E**<br>**0°C to 1000°C** | **R**<br>**−50°C to 250°C** | **T**<br>**0°C to 400°C** |
|---|---|---|---|---|---|
| $a_0$ | 0.0 | 0.0 | 0.0 | 0.0 | 0.0 |
| $a_1$ | $2.508355 \times 10^{-2}$ | $1.978425 \times 10^{-2}$ | $1.7057035 \times 10^{-2}$ | $1.8891380 \times 10^{-1}$ | $2.592800 \times 10^{-2}$ |
| $a_2$ | $7.860106 \times 10^{-8}$ | $-2.001204 \times 10^{-7}$ | $-2.3301759 \times 10^{-7}$ | $-9.3835290 \times 10^{-5}$ | $-7.602961 \times 10^{-7}$ |
| $a_3$ | $-2.503131 \times 10^{-10}$ | $1.036969 \times 10^{-11}$ | $6.5435585 \times 10^{-12}$ | $1.3068619 \times 10^{-7}$ | $4.637791 \times 10^{-11}$ |
| $a_4$ | $8.315270 \times 10^{-14}$ | $-2.549687 \times 10^{-16}$ | $-7.3562749 \times 10^{-17}$ | $-2.2703580 \times 10^{-10}$ | $-2.165394 \times 10^{-15}$ |
| $a_5$ | $-1.228034 \times 10^{-17}$ | $3.585153 \times 10^{-21}$ | $-1.7896001 \times 10^{-21}$ | $3.5145659 \times 10^{-13}$ | $6.048144 \times 10^{-20}$ |
| $a_6$ | $9.804036 \times 10^{-22}$ | $-5.344285 \times 10^{-26}$ | $8.4036165 \times 10^{-26}$ | $-3.8953900 \times 10^{-16}$ | $-7.293422 \times 10^{-25}$ |
| $a_7$ | $-4.413030 \times 10^{-26}$ | $5.099890 \times 10^{-31}$ | $-1.3735879 \times 10^{-30}$ | $2.8239471 \times 10^{-19}$ | – |
| $a_8$ | $1.057734 \times 10^{-30}$ | – | $1.0629823 \times 10^{-35}$ | $-1.2607281 \times 10^{-22}$ | – |
| $a_9$ | $-1.052755 \times 10^{-35}$ | – | $-3.2447087 \times 10^{-41}$ | $3.1353611 \times 10^{-26}$ | – |
| **Error** | ±0.05°C | ±0.05°C | ±0.02°C | ±0.02°C | ±0.03°C |

Temperature in °C, voltage in μV

the table is the error caused by the polynomial calculation and any error in the practical measurement is not considered.

In practice it can be very time consuming to evaluate polynomials with a very high degree. Polynomial calculation can be simplified and also speeded up if we re-write a polynomial as a sum of product terms. For example, consider the 4th order polynomial below:

$$T = a_0 + a_1 v + a_2 v^2 + a_3 v^3 + a_4 v^4 \tag{3-2}$$

We can re-write this polynomial as:

$$T = a_0 + v(a_1 + v(a_2 + v(a_3 + va_4))) \tag{3-3}$$

Equation 3-3 can be calculated by performing only multiplication and addition.

In some applications the temperature is known and we may want to calculate the thermocouple voltage. This can be done by using the inverse thermocouple polynomial:

$$V = c_0 + c_1 T + c_2 T^2 + \cdots + c_n T^n \tag{3-4}$$

where V is the thermocouple voltage, T is the temperature, and $c_0, c_1, c_2, \ldots, c_n$ are the thermocouple coefficients. The first ten thermocouple $c_i$ coefficients are shown in Table 3.4 for some popular thermocouples.

**Example 3.1**

The voltage measured at a K type thermocouple is 4096 $\mu$V. Calculate the temperature of the thermocouple.

**Solution 3.1**

Using equation 3-1 with 10 coefficients and re-writing this equation as addition and multiplication, we have:

$$T = a_0 + v(a_1 + v(a_2 + v(a_3 + v(a_4 + v(a_5 + v(a_6 + v(a_7 + v(a_8 + va_9)))))))) \tag{3-5}$$

Using the values of $a_0$ to $a_9$ from Table 3.3 we find the temperature as $T = 99.87°C$.

It is interesting to note that type K thermocouple generates a 4096 $\mu$V voltage at a temperature of +100°C. The error using the polynomial calculation in this example is therefore less than 0.13°C.

**Table 3.4** Thermocouple $c_i$ coefficients

| | **K** **0°C to 1372°** | **J** **−210°C to 760°C** | **E** **0°C to 1000°C** | **R** **−50°C to 1064°C** | **T** **0°C to 400°C** |
|---|---|---|---|---|---|
| $c_0$ | −17.600413686 | 0.0 | 0.0 | 0.0 | 0.0 |
| $c_1$ | 38.921204975 | 50.38118782 | 58.665508710 | 5.28961729765 | 38.748106364 |
| $c_2$ | $1.85587700 \times 10^{-2}$ | $3.047583693 \times 10^{-2}$ | $4.503227558 \times 10^{-2}$ | $1.3916658978 \times 10^{-2}$ | $3.32922279 \times 10^{-2}$ |
| $c_3$ | $-9.9457593 \times 10^{-5}$ | $-8.56810657 \times 10^{-5}$ | $2.890840721 \times 10^{-5}$ | $-2.388556930 \times 10^{-5}$ | $2.06182434 \times 10^{-4}$ |
| $c_4$ | $3.18409457 \times 10^{-7}$ | $1.322819530 \times 10^{-7}$ | $-3.30568967 \times 10^{-7}$ | $3.5691600106 \times 10^{-8}$ | $-2.18822568 \times 10^{-6}$ |
| $c_5$ | $-5.607284 \times 10^{-10}$ | $-1.7052958 \times 10^{-10}$ | $6.50244033 \times 10^{-10}$ | $-4.62347666 \times 10^{-11}$ | $1.09968809 \times 10^{-8}$ |
| $c_6$ | $5.6075059 \times 10^{-13}$ | $2.09480907 \times 10^{-13}$ | $-1.9197496 \times 10^{-13}$ | $5.007774410 \times 10^{-14}$ | $-3.0815759 \times 10^{-11}$ |
| $c_7$ | $-3.202072 \times 10^{-16}$ | $-1.2538395 \times 10^{-16}$ | $-1.2536600 \times 10^{-15}$ | $-3.73105886 \times 10^{-20}$ | $4.54791353 \times 10^{-14}$ |
| $c_8$ | $9.7151147 \times 10^{-20}$ | $1.56317257 \times 10^{-20}$ | $2.14892176 \times 10^{-18}$ | $1.577164824 \times 10^{-20}$ | $-2.7512902 \times 10^{-17}$ |
| $c_9$ | $-1.210472 \times 10^{-23}$ | - | $-1.4388042 \times 10^{-21}$ | $-2.81038625 \times 10^{-24}$ | - |

### 3.7.3 Using linear approximation

Another method of representing the thermocouple temperature voltage relationship is to use linear approximations over limited temperature ranges.

The linear approximation:

$$V = s \cdot T + b \tag{3-6}$$

where V is the thermocouple voltage, s is the slope, T is the temperature, and b is an offset voltage can be used to represent most thermocouples over limited temperature ranges. All thermocouples have the offset voltages equal to zero. The slope can be determined from the required operating range. Over the temperature range of 0°C to around +50°C we can use the averaged Seeback coefficients as the slope. The linearized equation then becomes:

$$V = s \cdot T \tag{3-7}$$

where V is the thermocouple voltage (in $\mu$V), s is the Seeback coefficient (in $\mu$V/°C), and T is the thermocouple junction temperature (in °C).

The error introduced by using equation 3-7 is only a few °C over the selected operating range and this error may be acceptable in many non-precision temperature measurement applications.

Table 3.5 gives the averaged Seeback coefficients for the popular thermocouple types over the operating range of 0°C to +50°C.

**Table 3.5** Seeback coefficients

| Type | $\mu$V/°C |
|---|---|
| K | 40.46 |
| J | 51.71 |
| T | 40.69 |
| E | 60.93 |
| B | 0.05 |
| S | 6.02 |
| R | 5.93 |

**Example 3.2**

Use the linearized equation to calculate the thermocouple voltage of a type K thermocouple when the junction temperature is +50°C and the thermocouple reference temperature is 0°C.

**Solution 3.2**

From equation 3-7,

$$V = s \cdot T$$

$$= 40.46 \times 50$$

$$= 2023\,\mu V$$

From thermocouple reference tables, the exact value is in fact 2023 μV.

**Example 3.3**

The junction voltage of a type K thermocouple is 960 μV. Calculate the junction temperature using equation 3-7 if the reference junction is at 0°C.

**Solution 3.3**

From equation 3-7,

$$T = V/s$$

$$= 960/40.46$$

$$= 23.72°C$$

From thermocouple reference tables, the exact value is 24°C, i.e. an error of 0.28°C.

## 3.8 The theory of the cold junction compensation

In this section we shall be looking at the cold junction compensation in more detail and derive some equations on how we can measure the temperature in practice.

Figure 3.8 shows a simple thermocouple measurement system where metal A and metal B are used in the construction of the thermocouple and the thermocouple is connected to a measurement device using a pair of copper wires.

Let $V_1$ be the voltage generated by the thermocouple junction. In addition to this, two more thermocouple voltages, named $V_2$ and $V_3$, are generated at the connection points with the copper.

Now consider Fig. 3.9, where $T_1$ is the temperature we wish to measure. Notice that an ice bath is used for the reference temperature and metal A is used in

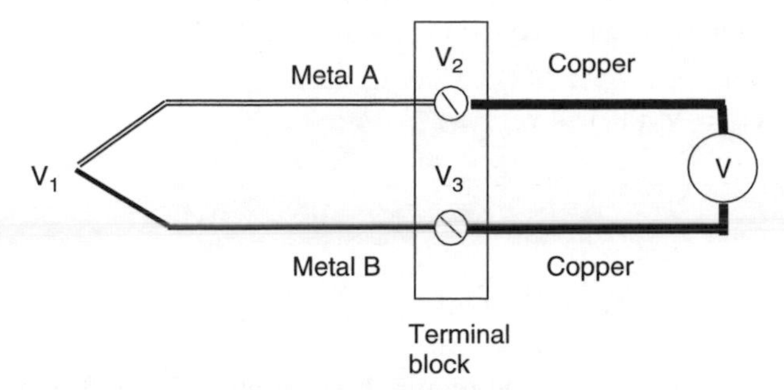


**Fig. 3.8** Simple thermocouple based measurement

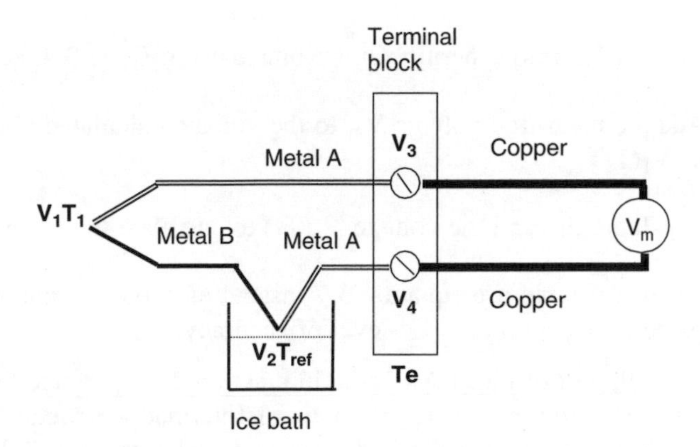


**Fig. 3.9** Using a thermocouple with the reference junction at 0°C

the second junction ($V_2$). The voltage measured by the measuring device can be written as:

$$V_m = V_1(T_1) + V_2(T_{ref}) + V_4(T_i) - V_3(T_i) \tag{3-8}$$

where $V_m$ is the measured voltage, $V_1(T_1)$ is the voltage generated by thermocouple 1 at temperature $T_1$, $V_2(T_{ref})$ is the voltage generated by thermocouple 2 at temperature $T_{ref}$, $V_3(T_i)$ is the voltage generated by thermocouple 3 at temperature $T_i$, and $V_4(T_i)$ is the voltage generated by thermocouple 4 at temperature $T_i$.

Junctions 3 and 4 are made of the same metals and $V_3 = V_4$. Equation 3-8 can then be written as:

$$V_m = V_1(T_1) + V_2(T_{ref}) \tag{3-9}$$

Junctions 1 and 2 are made of the same metals and connected in opposite polarities. Thus,

$$V_1(T_{ref}) = -V_2(T_{ref}) \tag{3-10}$$

We can now re-write equation 3-9 as:

$$V_m = V_1(T_1) - V_1(T_{ref}) \quad (3\text{-}11)$$

or,

$$V_1(T_1) = V_m + V_1(T_{ref}) \quad (3\text{-}12)$$

We can now calculate the required temperature $T_1$ as follows:

- Measure the reference temperature $T_{ref}$ (e.g. using a thermistor)
- Convert $T_{ref}$ into a thermocouple voltage using Table 3.4, i.e. calculate $V_1(T_{ref})$
- Add the measured voltage $V_m$ to the voltage calculated above. This will give us $V_1(T_1)$
- Use Table 3.3 and the voltage $V_1(T_1)$ to calculate the required temperature $T_1$

Notice that we can use equation 3-7 instead of Table 3.3 and Table 3.4 if we are prepared to accept a reduced level of accuracy.

In Fig. 3.9, part of metal A is used in junction 2. In practice this is not necessary and it can be removed when the Law of Intermediate Metals is considered (see Fig. 3.4). If we remove the intermediate metal from Fig. 3.9, the measurement circuit reduces to the one shown in Fig. 3.10. Notice that in this figure the junction where there is a connection to the copper wire (i.e. the reference junction) is isothermal, i.e. both connection points with the copper are kept at the same temperature, $T_{ref}$.

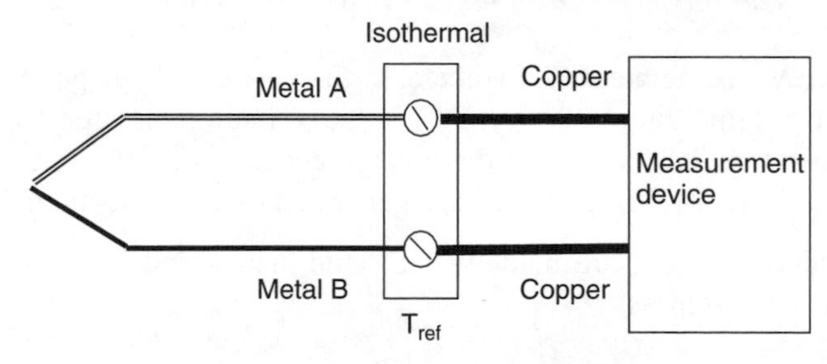


**Fig. 3.10** Practical measurement with a thermocouple

It is possible to compensate a thermocouple using a hardware compensation technique. Special ICs, such as the Linear Technology LT1025 (or Analog Devices AD594/AD595) can be used for this purpose. This chip has pins for various types of thermocouples. Figure 3.11 shows a typical application, using a type K thermocouple. In this circuit, the compensated thermocouple voltage is amplified to give an output voltage of 10 mV/°C and the gain is set to be 10 mV/40.46 μV = 247. The output of the operational amplifier is linearized as described in Section 3.7.3 and is accurate to a few degrees centigrade over the operating temperature. The operating temperature range of the standard LT1025 chip is 0°C to +70°C.

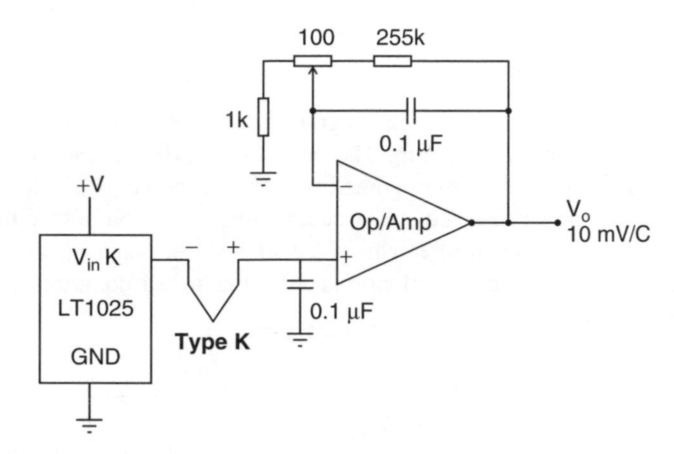


**Fig. 3.11** Using hardware to compensate a thermocouple

## 3.9 Microcontroller based practical thermocouple circuits

Thermocouple based temperature measurement requires two initial calculations: cold junction compensation and the calculation of the temperature using either the power series polynomials (or the linearized equation), or reference tables.

Cold junction compensation can be done either in software or in hardware. Figure 3.12 shows the block diagram where a temperature sensor (e.g. a thermistor, or a semiconductor temperature sensor) is used to measure the reference temperature and the cold junction compensation is done in software. An operational amplifier is used to increase the thermocouple signal value so that it can be digitized by the A/D converter. Both the digitized thermocouple voltage and the digitized reference voltages are fed to a microcontroller. The microcontroller calculates the measured temperature by applying cold junction compensation and the power series polynomial (or the linearized equation), as described in Section 3.8. The microcontroller drives a LCD display to show the measured temperature. In some noisy applications it may be necessary to use a low-pass filter to remove any electrical noise from the measurement system.

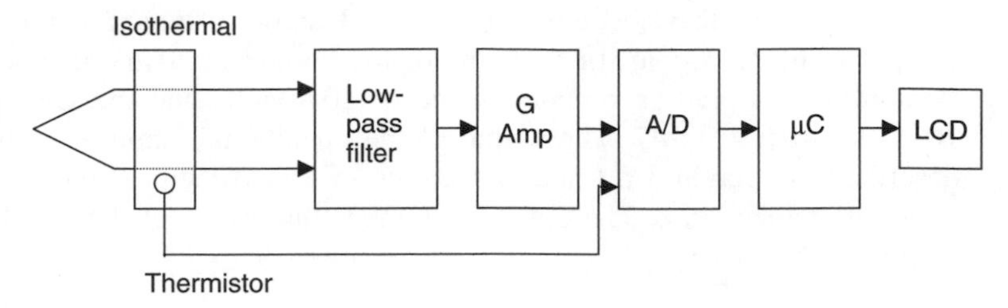


**Fig. 3.12** Software cold junction compensation and temperature measurement

Figure 3.13 shows the block diagram where the cold junction compensation is done using the LT1025 chip. The output of LT1025 is amplified and then fed to the A/D converter. It may sometimes be necessary to employ a low-pass filter to remove any unwanted noise at this stage. The output of the A/D converter is fed to a microcontroller which calculates and displays the temperature using either the power series polynomial, the linearized equation, or a thermocouple reference table.

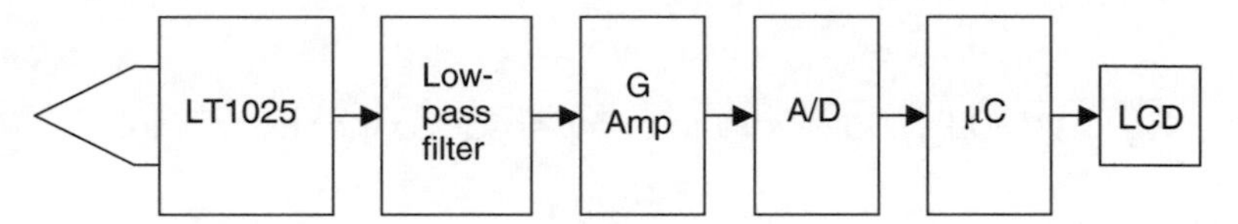


**Fig. 3.13** Hardware cold junction compensation and temperature measurement

# 3.10 PROJECT – Measuring temperature using a thermocouple and a microcontroller

In this section the design of a thermocouple based temperature measurement system will be described.

## 3.10.1 The specifications

Temperature sensor: Type K thermocouple

Temperature range: 0°C to 70°

Accuracy: 1°C

Compensation: Hardware

| | |
|---|---|
| Controller: | Microcontroller |
| Display: | LCD |
| Display format: | 6 characters, i.e. "nn.m C" |
| Update interval: | 1 second |

## 3.10.2 The hardware design

The block diagram of the thermocouple measurement system is shown in Fig. 3.14. Figure 3.15 shows the complete circuit diagram. Since the compensation will be done in hardware, a LT1025 type cold junction compensator chip is used. This chip has pins to connect type E, J, K, R, S, and T thermocouples and operates with a supply voltage from 4 V to 36 V. Typical supply current is

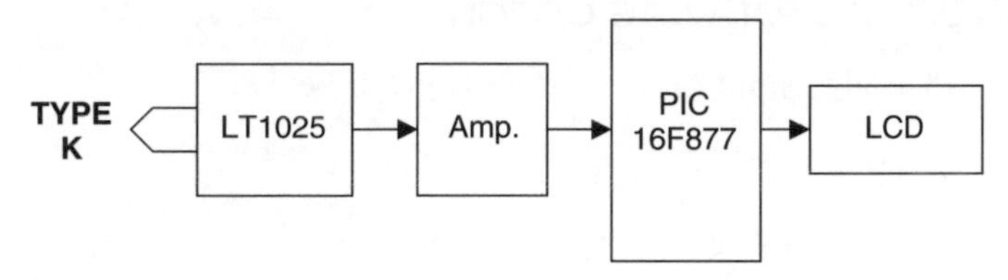


**Fig. 3.14** Block diagram of the system

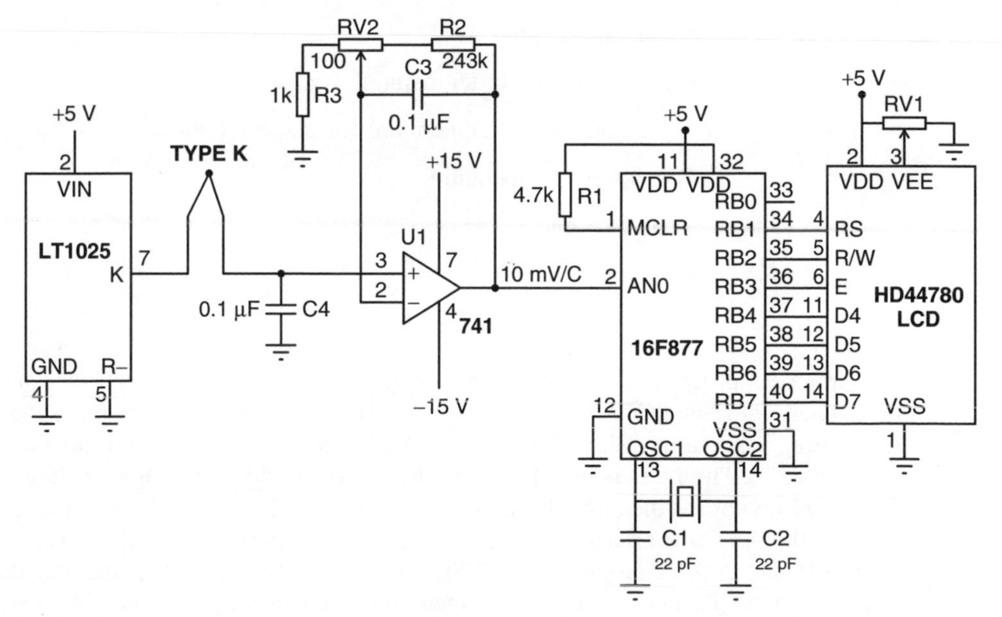


**Fig. 3.15** Circuit diagram of the system

80 μA, resulting in less than 0.1°C internal temperature rise for supply voltages under 10 V. A 5 V supply is thus chosen for this chip. The output of the chip (pin K) is amplified such that the output voltage is 10 mV/°C. Linearized equation shall be used and thus the required amplifier gain is approximately 10 mV/40.46 μV = 247. Two fixed resistors and a variable resistor are used to adjust the gain appropriately. It is important that the resistors should be chosen to have very small tolerances for accuracy. The amplifier should be adjusted to eliminate any offset errors. It is also recommended to use a low-pass filter in noisy environments. The output of the amplifier is connected to one of the analogue inputs (AN0) of the PIC16F877 microcontroller. The A/D has a 10-bit resolution and assuming a 5 V reference supply, one LSB of the converter is equivalent to 5000/1024 = 4.88 mV. Thus, the measurement accuracy will be about 0.5°C. A 4 MHz crystal is used as the clock source. Port B of the microcontroller is connected to a single line HD44780 type LCD display where the temperature is displayed every second in the "nn.mm C" format (e.g. 20.50°C).

### 3.10.3 The software design

The PDL algorithm of the system is given below:

```
BEGIN
        Initialize the LCD
        Initialize the microcontroller
        DO FOREVER
                Read analog input AN0
                Convert reading to temperature
                Scale and format temperature for display ("nn.m C")
                Display the temperature
                Wait 1 second
        ENDDO
END
```

The complete program listing is given in Fig. 3.16. The LCD is connected to Port B and constant *LCDPORT* is set accordingly. Variable *LSB* is equal to the value of an LSB in mV (i.e. 5000/1024). Variable *temp* is defined as a floating point variable. The LCD is initialized and the cursor is set to the home position. The A/D converter is then initialized by configuring the top 6-bits of data input to be zero (the A/D converter is 10-bits). The A/D output is stored in PIC registers ADRESH and ADRESL where ADRESH stores the two top bits of the 10-bit converted data. This converted data is then combined into a 10-bit number and stored in variable *temp*. The integer and the fractional parts of the temperature

```
/*********************************************************************
*
*
*      PROJECT:     PROJECT3
*      FILE:        PROJ3.C
*      DATE:        November 2001
*      PROCESSOR:   PIC16F877
*      COMPILER:    FED C
*
*
* This project measures the temperature with a type K thermocouple and
* then displays on a LCD every second in the format "nn.m C".
*
*********************************************************************/

#include <P16F877.h>
#include <displays.h>
#include <delays.h>
#include <strings.h>
#include <maths.h>

const int LCDPORT=&PORTB;                    //Define LCD port as Port B


void main()
{
            float temp,LSB;
            int msd,lsd;
            char temperature[6];

            LSB = 5000.0/1024.0;                   //LSB in mV

            /* Initialize the LCD */
            LCD(-1);                         //Init. LCD to 1 line
            LCD(257);                        //Clear LCD and home

            while(1)                         //DO FOREVER
            {
                 /* Configure the A/D */
                 ADCON1 = 0x80;                   //Select 10 bits
                 ADCON0 = 0x41;

                 /* Start A/D conversion */
                 ADCON0=0x45;                     //Start A/D conversion

                 while((ADCON0 & 4) !=  0); //Wait for conversion
                 temp=256.0*(float)ADRESH+(float)ADRESL;  //Get temperature

                 /* Convert reading to temperature */
                 temp = temp*LSB;                 //temp in mV
                 temp = temp/10.0;                //temp in degrees C

                 /* Format for the display */
                 LCD(257);                        //Clear LCD and home

                 msd=temp;                        //msd digit
                 lsd=10.0*(temp-msd);             //lsd digit

                 iPrtString(temperature,msd); //msd
```

**Fig. 3.16** Program listing of the project

```
                /* Insert ".", lsd digit, and "C" character */
                temperature[2]='.';
                cPrtString(temperature+3,lsd);
                temperature[4]=' ';
                temperature[5]='C';

                /* Display the temperature as "nn.m C" */
                LCDString(temperature);

                Wait(1000);
        }                                           //ENDDO
}
```

**Fig. 3.16** *(Continued)*

are then extracted and stored in variables *msd* and *lsd* respectively. These values are then converted to string format and stored in a string called *temperature*. The decimal point and the Celsius indicator are then added to this string and then the string data is sent to the LCD. The process is repeated continuously with one second delay between each output.

## 3.11 Exercises

1. What is a thermocouple reference junction?

2. If the thermocouple hot junction is at 120°C and the reference junction is at 30°C what will be the measured temperature without any cold junction compensation?

3. Explain the Law of Intermediate Metals. How can this law be applied to a thermocouple measurement system?

4. What is a thermocouple extension cable? Explain why it is important to use the correct extension cable.

5. The voltage measured using a type K thermocouple is 2040 μV. Calculate the temperature of the thermocouple using the power series polynomials.

6. The temperature at the junction of a type J thermocouple is 37°C. Use the power series polynomials to calculate the thermocouple voltage in μV.

7. Use the linearized equation to calculate the hot junction temperature of a type K thermocouple when the thermocouple voltage is 1612 μV and the thermocouple reference temperature is 0°C.

8. Use the power series polynomials to calculate the hot junction temperature in exercise 7.

9. Use the thermocouple reference tables to find the hot junction temperature in exercise 7. What is the error when the linearized equation is used? Is this error acceptable? Is the error less when the power series polynomial is used? Which method would you choose? Why?

10. Explain how the cold junction compensation can be achieved both in software and in hardware. Which method would you choose? Why?

11. Modify the program code given in Project 3.10 so that the thermocouple temperature resistance relationship is read from a look-up table. Would you prefer this method?

12. Modify the circuit diagram of Project 3.10 so that a thermistor can be used to measure the reference temperature.

13. Modify the program code given in Project 3.10 so that the cold junction compensation is done in software. Assume that a thermistor is available to measure the reference temperature (refer to Chapter 5 on thermistors).

# Chapter 4

# RTD Temperature Sensors

## 4.1 RTD principles

The term RTD is short for Resistance Temperature Detector. The RTD is a temperature sensing device whose resistance increases with temperature. RTDs are quite linear devices and a typical temperature-resistance characteristic is shown in Fig. 4.1. RTDs operates on the principle that the electrical resistance of metals change with temperature. Although in theory any kind of metal can be used for temperature sensing, in practice metals with high melting points which can withstand the effects of corrosion, and those with high resistivities are chosen. Table 4.1 gives the resistivities of some commonly used RTD metals. Gold and silver have low resistivities and as a result their resistances are relatively low, making the measurement difficult. Copper has low resistivity but is sometimes used because of its low cost. The most commonly used RTDs are made of either nickel, platinum, or nickel alloys. Nickel sensors are used in cost sensitive applications such as consumer goods and they have a limited temperature range. Nickel alloys, such as nickel-iron is lower in cost than the pure nickel and in addition it has a higher operating temperature. Platinum is by far the most common RTD material, mainly because of its high resistivity and long term stability in air.

RTDs have excellent accuracies over a wide temperature range and some RTDs have accuracies better than 0.001°C. Another advantage of the RTDs is that they drift less than 0.1°C/year.

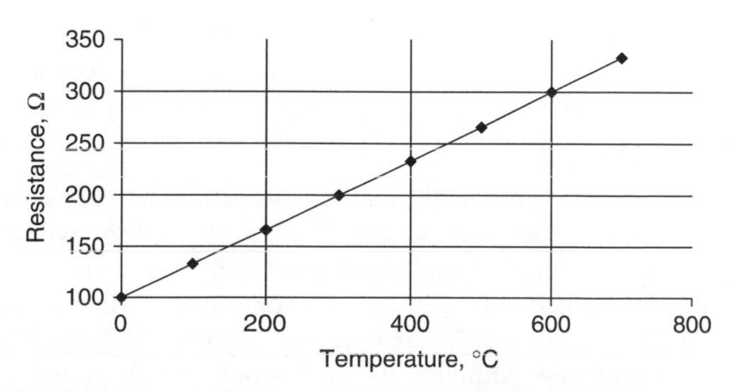


**Fig. 4.1** Typical RTD temperature-resistance characteristics

**Table 4.1** Resistivities and temperature ranges of some RTD metals

| Metal | Resistivity (ohm/cmf) |
|---|---|
| Silver | 8.8 |
| Copper | 9.26 |
| Gold | 13.00 |
| Tungsten | 30.00 |
| Nickel | 36.00 |
| Platinum | 59.00 |

cmf = circular mil foot

RTDs are difficult to measure because of their low resistances and only slight changes with temperature, usually in the order of 0.40/°C. To accurately measure such small changes in resistance, special circuit configurations are usually needed. For example, long leads could cause errors as it introduces extra resistance to the circuit.

RTDs are resistive devices and a current must pass through the device so that the voltage across the device can be measured. This current can cause the RTD to self-heat and consequently it can introduce errors into the measurement. Self-heat can be minimized by using the smallest possible excitation current. The amount of self-heat also depends on where and how the sensor is used. An RTD can self-heat much quicker in still air than in a moving liquid.

## 4.2 RTD types

In order to achieve high stability and accuracy, RTD sensors must be contamination free. Below about 250°C the contamination is not much of a problem, but above this temperature, special manufacturing techniques are used to minimize the contamination of the RTD element.

The RTD sensors are usually manufactured in two forms: wire wound, or thin film. Figure 4.2 shows a typical RTD sensor. Wire wound RTDs are made by winding a very fine strand of platinum wire into a coil shape around a non-conducting material (e.g. ceramic or glass) until the required resistance is obtained. The assembly is then treated to protect short-circuit and to provide vibration resistance. Although the wire wound RTDs are very stable, the thermal contact between the platinum and the measured point is not very good and results

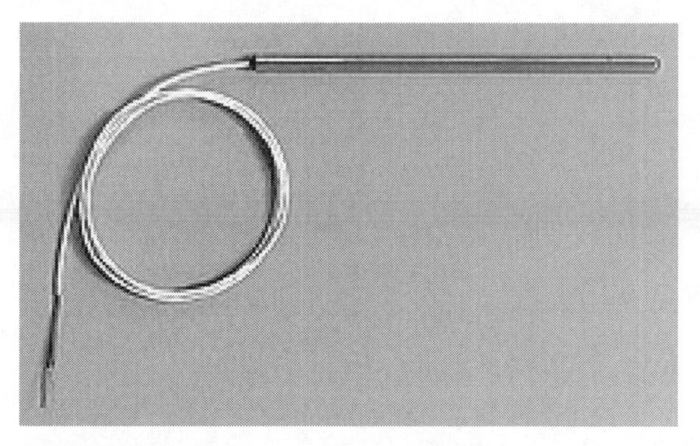

**Fig. 4.2** A typical RTD sensor

in slow thermal response. Thin film RTDs are made by depositing a layer of platinum in a resistance pattern on a ceramic substrate. The film is treated to have the required resistance and then coated with glass or epoxy for moisture resistance and to provide vibration resistance. Thin film RTDs have the advantages that they provide a fast thermal response, are less sensitive to vibration, and they cost less than their wire wound counterparts. Thin film RTDs can also provide a higher resistance for a given size. These RTDs are less stable than the wire wound ones but they are becoming very popular as a result of their considerably lower costs.

## 4.3 RTD temperature resistance relationship

Every metal has a resistance and this resistance is directly proportional to length and the resistivity of the metal, and inversely proportional to its cross-sectional area:

$$R = \rho L/A \tag{4-1}$$

where R is the resistance of the metal, $\rho$ is the resistivity, L is the length of the metal, and A is the cross-sectional area of the metal.

The resistivity increases with increasing temperature as shown in equation 4-2:

$$\rho_t = \rho_0[1 + a(t - t_0)] \tag{4-2}$$

where $\rho_t$ is the resistivity at temperature t, $\rho_0$ is the resistivity at a standard temperature $t_0$, and a is the temperature coefficient of resistance.

Setting $t_0$ to 0°C, we can re-write equation 4-2 as:

$$\rho_t = \rho_0(1 + a \cdot t) \tag{4-3}$$

If $R_0$ is the resistance at 0°C and $R_t$ is the resistance at temperature t, we can re-write equation 4-1 as:

$$R_0 = \rho_0 L/A \tag{4-4}$$

and

$$R_t = \rho_t L/A \tag{4-5}$$

or, from equations 4-3 and 4-5,

$$R_t = \rho_0(1 + a \cdot t)L/A \tag{4-6}$$

and using equations 4-4 and 4-6, we can write the relationship between the temperature and the resistance as:

$$R_t = R_0(1 + a \cdot t) \tag{4-7}$$

Equation 4-7 is a simplified model of the RTD temperature-resistance relationship. In practice, temperature-resistance relationship of the RTDs are approximated by an equation known as the *Callendar–Van Dusen* equation which gives very accurate results. This equation has the form:

$$R_t = R_0[1 + At + Bt^2 + C(t - 100)^3] \tag{4-8}$$

where A, B, and C are constants which depend upon the material. Above 0°C, the constant C is equal to zero and we can re-write equation 4-8 as:

$$R_t = R_0[1 + At + Bt^2] \tag{4-9}$$

Thus, if we know the constants A and B and the resistance at 0°C then we can calculate the resistance at any other positive temperature using equation 4-9. However, in practice it is required to calculate the temperature from a knowledge of the RTD resistance. Equation 4-9 is a quadratic equation in t and it can be solved to give:

$$t = \frac{-R_0A + \sqrt{R_0{}^2A^2 - 4R_0B(R_0 - R_t)}}{2R_0B} \tag{4-10}$$

An example is given below to illustrate the steps in calculating the temperature.

**Example 4.1**

The resistance of an RTD is 100 Ω at 0°C. If the A and B constants are as given below, calculate the temperature when the resistance is measured as 138 Ω.

Given:

$$A = 3.908 \times 10^{-1}$$

$$B = -5.775 \times 10^{-7}$$

$$R_0 = 100\,\Omega$$

$$R_t = 138\,\Omega$$

Required:

$$t = ?$$

**Solution 4.1**

The temperature can be calculated using equation 4-10.

From equation 4-10,

$$t = \frac{-100 \times 3.908 \times 10^{-1} + \sqrt{100^2 \times (3.908 \times 10^{-1})^2 - 4 \times 100 \times 5.775 \times 10^{-7} \times 38}}{-2 \times 100 \times 5.775 \times 10^{-7}}$$

which gives,

$$t = 99.8^\circ C$$

Some manufacturers provide RTD constants known as $\alpha$, $\beta$, and $\delta$. Knowing these constants, we can calculate the standard A, B, and C constants as:

$$A = \alpha + \frac{\alpha \cdot \delta}{100} \tag{4-11}$$

$$B = \frac{-\alpha \cdot \delta}{100^2} \tag{4-12}$$

$$C = \frac{-\alpha \cdot \beta}{100^4} \quad \text{only for } t < 0 \tag{4-13}$$

Parameter $\alpha$ is important as it is used in defining the RTD standards. This parameter is the change in RTD resistance from 0°C to 100°C, divided by the resistance at zero degrees Celsius, divided by 100°C. Thus, $\alpha$ represents the mean resistance change referred to the nominal resistance at 0°C:

$$\alpha = \frac{R_{100} - R_0}{R_0 \cdot 100^\circ C} \tag{4-14}$$

**Example 4.2**

The $\alpha$, $\beta$, and $\delta$ constants of a platinum RTD used to measure the temperature between 0°C and 80°C are specified by the manufacturers as:

$$\alpha = 0.003850$$

$$\beta = 0.10863$$

$$\delta = 1.4999$$

Calculate the standard RTD constants A, B, and C.

**Solution 4.2**

From equations 4-11 and 4-12,

$$A = \alpha + \frac{\alpha \cdot \delta}{100} = 0.003850 + \frac{0.003850 \times 1.4999}{100}$$

or,

$$A = 3.9083 \times 10^{-3}$$

and,

$$B = \frac{-\alpha \cdot \delta}{100^2} = \frac{-0.003850 \times 1.4999}{100^2}$$

or,

$$B = -5.775 \times 10^{-7}$$

C = 0 since the RTD is used to measure positive temperatures

Appendix A gives a table of the platinum RTD temperature-resistance characteristics for the temperature range of 0°C to 100°C.

## 4.4 RTD standards

Platinum RTDs conform to either IEC/DIN standard or reference-grade standards. The main difference is in the purity of the platinum used. The IEC/DIN standard defines pure platinum which is contaminated with other platinum group metals. The reference-grade platinum is made from 99.99% pure platinum.

The most commonly used international RTD standard is the IEC 751. Other international standards such as BS 1904 and DIN 43760 match this standard. IEC 751 is based on platinum RTDs with a resistance of 100 Ω at 0°C and a parameter of 0.00385. Two performance classes are defined under the IEC 751: Class A and Class B. These performance classes, also known as DIN A and DIN B (due to DIN 43760) define tolerances on ice point and temperature accuracy.

**Table 4.2** IEC 751 RTD class properties

| Parameter | IEC 751 Class A | IEC 751 Class B |
|---|---|---|
| $R_0$ | 100 Ω ± 0.06% | 100 Ω ± 0.12% |
| Alpha, α | 0.00385 ± 0.000063 | 0.00385 ± 0.000063 |
| Range | −200°C to 650°C | −200°C to 850°C |
| Temp. tolerance | ±(0.15 + 0.002\|T\|)°C | ±(0.3 + 0.005\|T\|)°C |

**Table 4.3** Temperature tolerances of IEC 751 RTD sensors

| Temperature °C | IEC 751 Tolerance | |
|---|---|---|
| | Class A | Class B |
| | ±°C | ±°C |
| −200 | 0.55 | 1.3 |
| −100 | 0.35 | 0.8 |
| 0 | 0.15 | 0.3 |
| 100 | 0.35 | 0.8 |
| 200 | 0.55 | 1.3 |
| 300 | 0.75 | 1.8 |
| 400 | 0.95 | 2.3 |
| 500 | 1.15 | 2.8 |
| 600 | 1.35 | 3.3 |
| 650 | 1.45 | 3.6 |
| 700 | – | 3.8 |
| 800 | – | 4.3 |
| 850 | – | 4.6 |

**Table 4.4** Resistance tolerances of IEC 751 RTD sensors

| Temperature °C | IEC 751 Tolerance | |
|---|---|---|
| | Class A | Class B |
| | ±°Ω | ±°Ω |
| −200 | 0.24 | 0.56 |
| −100 | 0.14 | 0.32 |
| 0 | 0.06 | 0.12 |
| 100 | 0.13 | 0.30 |
| 200 | 0.20 | 0.48 |
| 300 | 0.27 | 0.64 |
| 400 | 0.33 | 0.79 |
| 500 | 0.38 | 0.93 |
| 600 | 0.43 | 1.06 |
| 650 | 0.46 | 1.13 |
| 700 | – | 1.17 |
| 800 | – | 1.28 |
| 850 | – | 1.34 |

The Callendar–Van Dusen parameters of IEC 751 standard RTDs are as follows:

$$A = 3.9083 \times 10^{-3}$$

$$B = -5.775 \times 10^{-7}$$

$$C = -4.183 \times 10^{-12}$$

Table 4.2 lists the properties of both classes. The temperature and resistance tolerances are listed in Tables 4.3 and 4.4, respectively. These tolerances are also plotted in Fig. 4.3 and Fig. 4.4.

## 4.4.1 Class A standard

This standard defines the temperature within the range −200°C to +650°C. The resistance at 0°C is $R_0 = 100\,\Omega$, and at 100°C it is $R_{100} = 138.5\,\Omega$. Class A

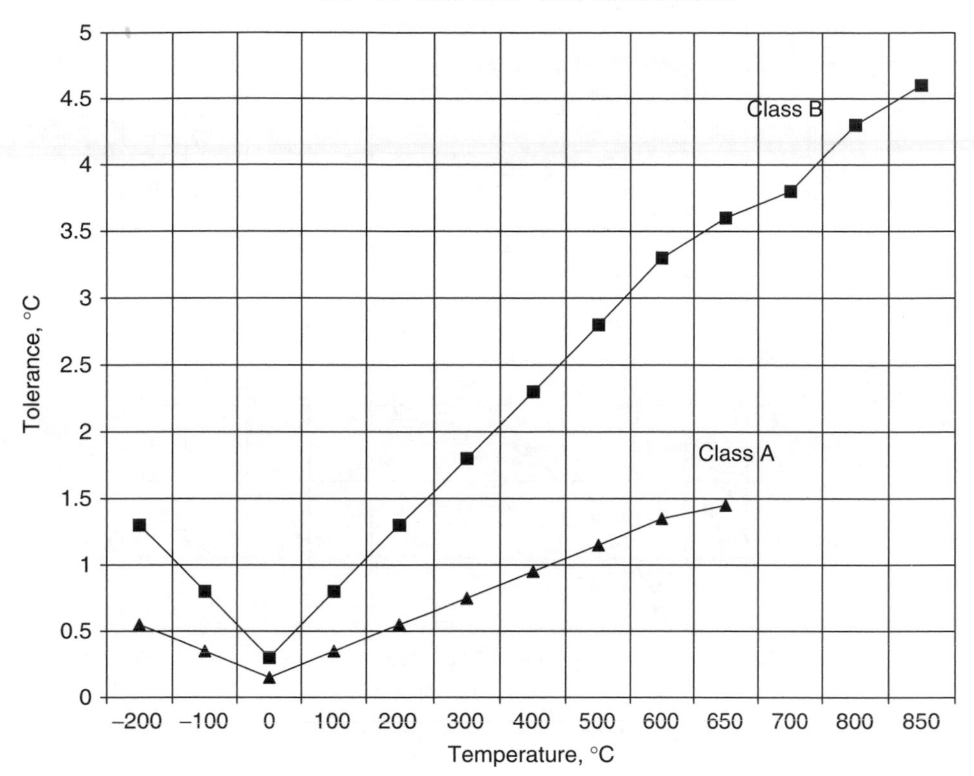


**Fig. 4.3** IEC 751 RTD temperature tolerances

temperature tolerance is:

$$dt = \pm(0.15 + 0.002 \cdot |t|)^{\circ}C$$

where |t| is the absolute value of the temperature in °C, e.g. at 0°C the temperature tolerance is ±0.15°C.

## 4.4.2 Class B standard

Class B provides less accuracy than Class A. This standard defines the temperature within the range −200°C to +850°C. The resistance at 0°C is $R_0 = 100\,\Omega$, and at 100°C it is $R_{100} = 138.5\,\Omega$. Class B temperature tolerance is:

$$dt = \pm(0.3 + 0.005 \cdot |t|)^{\circ}C$$

e.g. at 0°C the temperature tolerance is ±0.3°C.

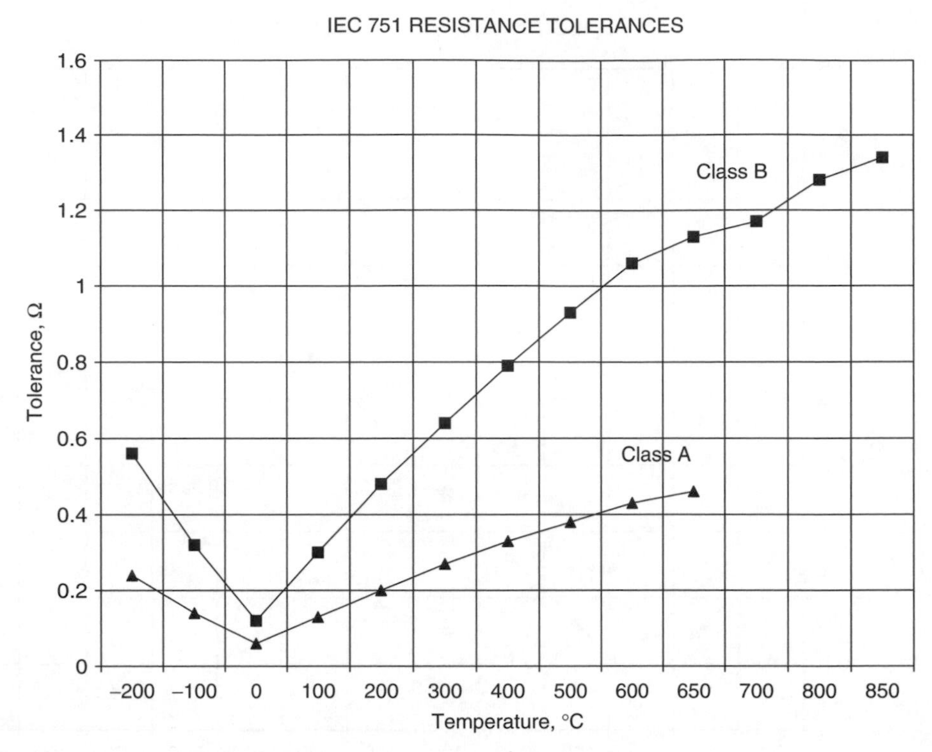


**Fig. 4.4** IEC 751 RTD resistance tolerances

# 4.5 Practical RTD circuits

Platinum RTDs are very low resistance devices and they produce very little resistance changes for large temperature changes. For example, a 1°C temperature change will cause a 0.384 Ω change in resistance, so even a small error in measurement of the resistance can cause a large error in the measurement of the temperature. For example, consider a 100 Ω RTD with a 1 mA excitation current. If the temperature rises by 1°C, the voltage across the RTD will increase by only about 0.5 mV. If the excitation current is 100 mA, the same change in temperature will result in a 50 mV change in the voltage, which is much easier to measure accurately. A high excitation current however should be avoided since it could give rise to self-heating of the sensor. The resistance of the wires leading to the sensor could give rise to errors when long wires are used. RTDs are usually used in bridge circuits for precision temperature measurement applications. Various practical RTD circuits are given in this section.

## 4.5.1 Simple current source circuit

Figure 4.5 shows a simple RTD circuit where a constant current source I is used to pass a current through the RTD. The voltage across the RTD is measured and

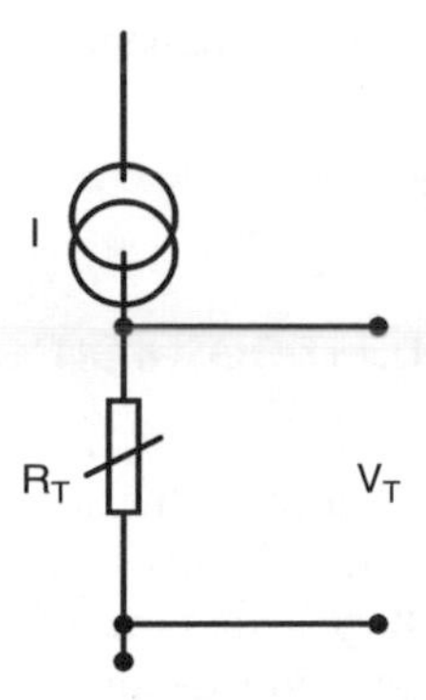


**Fig. 4.5** Simple current source RTD circuit

then the resistance of the RTD is calculated. The temperature can then be found by using equation 4-10. This circuit has the disadvantage that the resistance of the wires could add to the measured resistance and hence it could cause errors in the measurement. For example, a lead resistance of 0.3 Ω in each wire adds 0.6 Ω error to the resistance measurement. For a platinum RTD with a = 0.00385, this resistance is equal to 0.6 Ω/(0.385 Ω/°C) = 1.6°C error. Care should also be taken not to pass a large current through the RTD since this can cause self-heating of the RTD and hence change of its resistance. For example, a current of 1 mA through a 100 Ω RTD will generate 100 μW of heat. If the sensor element is unable to dissipate this heat, it will cause the resistance of the element to increase and hence an artificially high temperature will be reported.

## 4.5.2 Simple voltage source circuit

Figure 4.6 shows how an excitation current can be passed through an RTD by using a constant voltage source. This circuit suffers from the same problems as the one in Fig. 4.5. The voltage across the RTD element is:

$$V_T = V_S \cdot \frac{R_T}{R_T + R_S} \tag{4-15}$$

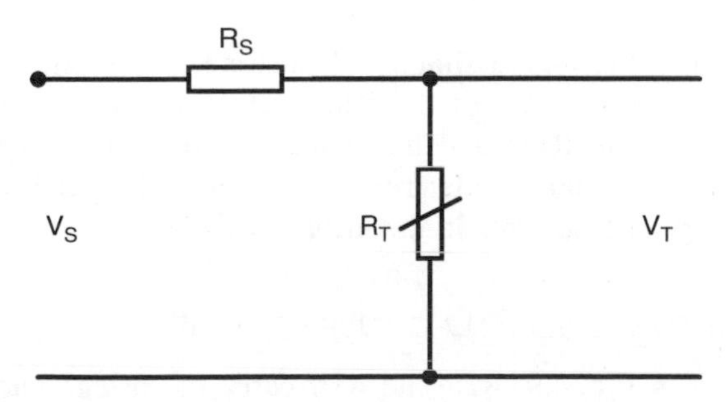


**Fig. 4.6** Simple voltage source RTD circuit

and the resistance of the RTD element can be calculated as:

$$R_T = R_S \cdot \frac{V_T}{V_S - V_T} \tag{4-16}$$

### 4.5.3 Four-wire RTD measurement

When long leads are used (e.g. greater than about 5 metres), it is necessary to compensate for the resistance of the lead wires. In the four-wire measurement method (see Fig. 4.7), one pair of wires carry the current through the RTD, the other pair senses the voltage across the RTD. In Fig. 4.7, $RL_1$ and $RL_4$ are the lead wires carrying the current and $RL_2$ and $RL_3$ are the lead wires for measuring the voltage across the RTD. Since only negligible current flows through the sensing wires (voltage measurement device having a very high internal resistance), the lead resistances $RL_2$ and $RL_3$ can be neglected. Four-wire RTD measurement gives very accurate results and is the preferred method for accurate, precision RTD temperature measurement applications.

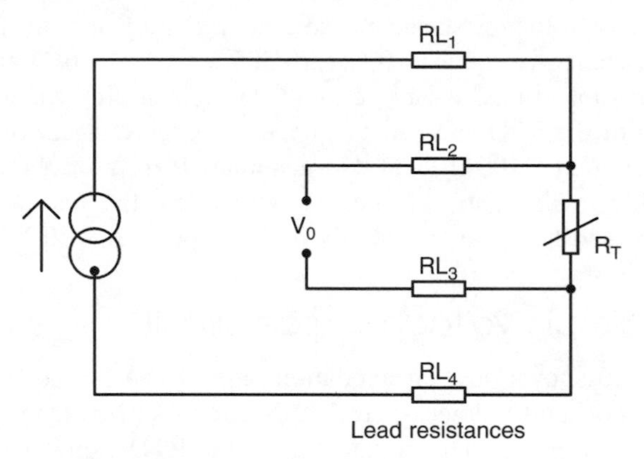


**Fig. 4.7** Four-wire RTD circuit

### 4.5.4 Simple RTD bridge circuit

As shown in Fig. 4.8, a simple Wheatstone bridge circuit can be used with the RTD at one of the legs of the bridge. As the temperature changes so does the resistance of the RTD and the output voltage of the bridge. This circuit has the disadvantage that the lead resistances $RL_1$ and $RL_2$ add to the resistance of the RTD, giving an error in the measurement.

### 4.5.5 Three-wire RTD bridge circuit

As shown in Fig. 4.9, $RL_1$ and $RL_3$ carry the bridge current. When the bridge is balanced, no current flows through $RL_2$ and thus the lead resistance $RL_2$ does

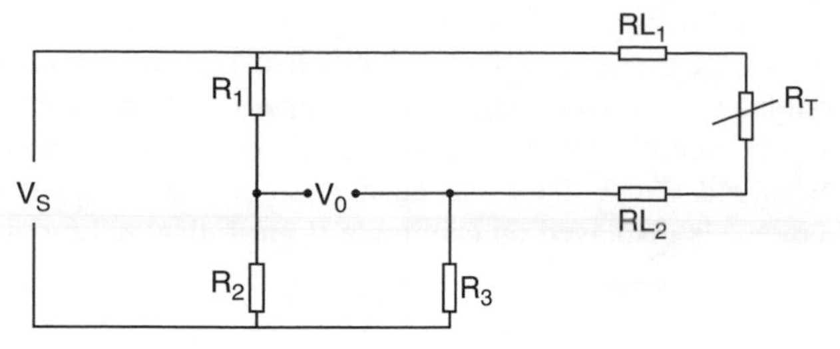


**Fig. 4.8** Simple RTD bridge circuit

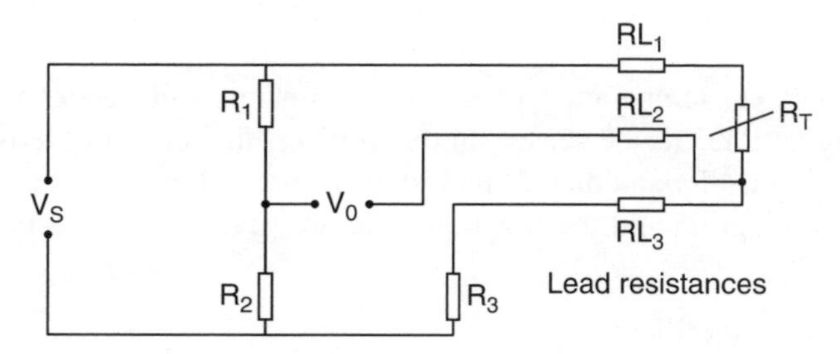


**Fig. 4.9** Three-Wire RTD bridge circuit

not introduce any errors into the measurement. The effects of $RL_1$ and $RL_3$ at different legs of the bridge cancel out since they have the same lengths and are made up of the same material.

# 4.6 Microcontroller based RTD temperature measurement

RTDs give analogue output voltages and the measured temperature can simply be displayed using an analogue meter (e.g. a voltmeter). Although this may be acceptable for some applications, accurate temperature measurement and display require digital techniques.

Figure 4.10 shows how the temperature can be measured using an RTD and a microcontroller. The temperature is sensed by the RTD and a voltage is produced which is proportional to the measured temperature. This voltage is filtered using

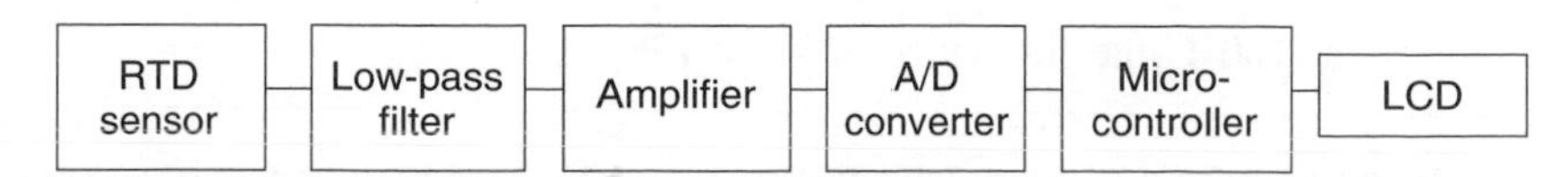


**Fig. 4.10** Microcontroller based RTD temperature measurement

a low-pass filter to remove any unwanted high frequency noise components. The output of the filter is then amplified using an operational amplifier so that this output can be fed to an A/D converter. The digitized voltage is then read by the microcontroller and the resistance of the RTD is calculated. After this, the measured temperature can be calculated using equation 4-10. Finally, the temperature is displayed on a suitable digital display (e.g. an LCD).

## 4.7 PROJECT – Designing a microcontroller based temperature measurement system using an RTD

A complete temperature measurement system will be described in this section using an RTD as the sensor. In this project, the temperature is calculated using a microcontroller and then displayed every second on a LCD display. The complete circuit diagram and the program code are given in this section.

### 4.7.1 Specifications

The system is required to have the following specifications:

| | |
|---|---|
| Temperature range: | 0°C to +99°C, accurate to a few °C |
| Sensor: | Platinum RTD |
| Display type: | LCD |
| Display format: | 6 characters, displayed as “nn.m C” |
| Display update: | 1 second |
| System: | Microcontroller based |

### 4.7.2 Design

RTDs from many manufacturers can be chosen for this project and a standard Class B RTD is chosen with an $\alpha$ of 0.00385.

We are assuming that the RTD is close to the microcontroller and the lead resistance errors are negligible. A simple constant voltage source type design (Fig. 4.6) will be used to sense the temperature.

From the platinum RTD characteristics,

at 0°C the resistance is $R_T = 100\,\Omega$

at 100°C the resistance is $R_T = 138.5\,\Omega$

If we assume a source resistance of $R_S = 1k$, and a source voltage of $V_S = 5\,V$, the voltage across the RTD element at either end of the operating temperature will be:

At 0°C, $V_T = 454\,\text{mV}$

At 100°C, $V_T = 608\,\text{mV}$

If we use an operational amplifier with a gain of 5, the voltage range seen by the A/D converter will be $5 \times 454 = 2270\,\text{mV}$ to $5 \times 608 = 3040\,\text{mV}$. Assuming that a 10-bit A/D converter will be used (e.g. the PIC16F877 microcontroller) with a full-scale of 5 V, 1 LSB $= 5000/1024 = 4.88\,\text{mV}$ but the input voltage range is $3040 - 2270 = 770\,\text{mV}$, or $770\,\text{mV}/100 = 7.70\,\text{mV}/°\text{C}$. Thus, our system will be accurate to about 1°C.

## 4.7.3 The circuit diagram

The block diagram of the project is shown in Fig. 4.11. The complete circuit diagram is shown in Fig. 4.12. A standard RTD is connected to a +5 V constant voltage source. The voltage across the RTD is fed to an operational amplifier with a gain of 5. Low-pass filter circuit is not used here for simplicity. The

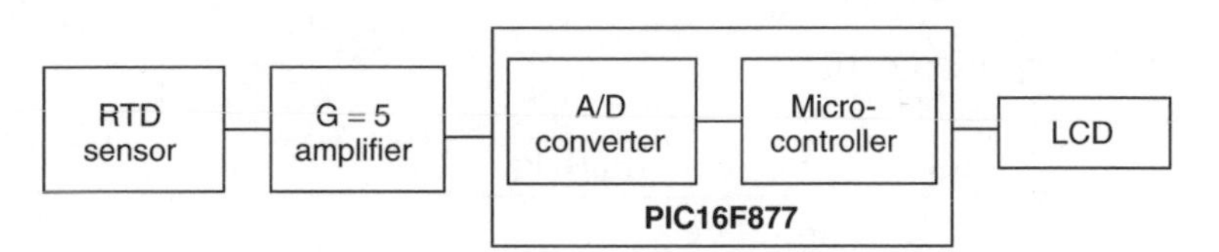


**Fig. 4.11** Block diagram of the temperature measurement system

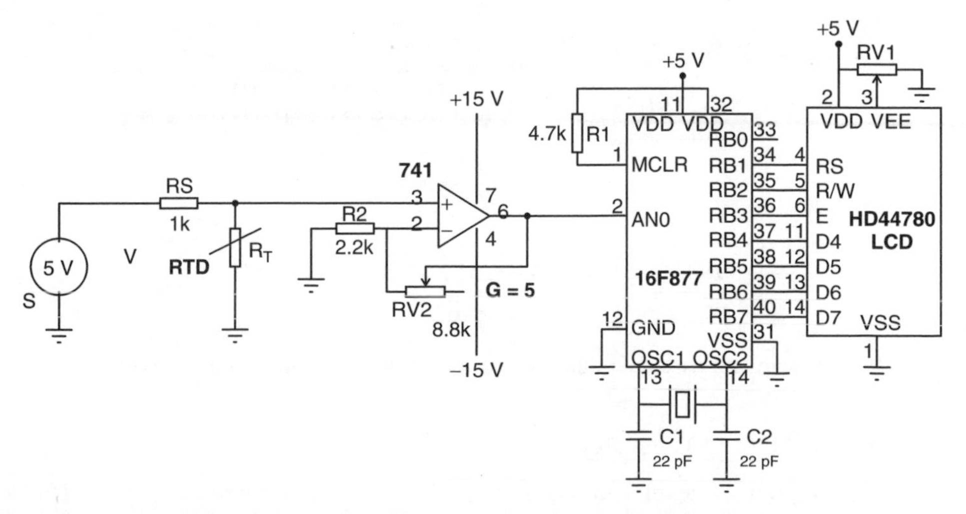


**Fig. 4.12** Circuit diagram of the temperature measurement system

output of the operational amplifier is sent to one of the A/D converter channels of a PIC16F877 microcontroller which is operated from a 4 MHz crystal clock source. A parallel LCD is connected to port B of the microcontroller where the measured temperature is displayed continuously in °C.

## 4.7.4 Operation of the circuit

The voltage across the RTD is converted into digital form and the resistance of the RTD ($R_T$) is calculated using equation 4-16. The temperature is then calculated using equation 4-10.

From equation 4-16,

$$R_T = \frac{V_T \cdot R_S}{V_S - V_T} \tag{4-17}$$

where $V_T$ is the voltage across the RTD, $V_S = 5\,V$, and $R_S = 1k$.

Thus,

$$R_T = \frac{10^3 \cdot V_T}{5 - V_T}\,\Omega$$

From equation 4-10,

$$t = \frac{-R_0A + \sqrt{R_0{}^2A^2 - 4R_0B(R_0 - R_t)}}{2R_0B} \tag{4-10}$$

where t is the measured temperature (°C), $R_t$ is the resistance at temperature t, $R_0 = 100\,\Omega$, $A = 3.9083 \times 10^{-3}$, and $B = -5.775 \times 10^{-7}$

Thus,

$$t = \frac{-100 \times 3.9083 \times 10^{-3} + \sqrt{\begin{array}{r}100^2 \times (3.9083 \times 10^{-3})^2 + 4 \times 100 \\ \times 5.775 \times 10^{-7}(100 - R_t)\end{array}}}{-2 \times 100 \times 5.775 \times 10^{-7}}$$

or,

$$t = \frac{-0.39083 + \sqrt{0.15274 - 2310 \times (R_t - 100) \times 10^{-7}}}{-1155 \times 10^{-7}} \tag{4-18}$$

If $R_t$ is known, we can calculate the temperature from equation 4-18.

## 4.7.5 Program listing

The program is developed using the FED C compiler, targeted to a PIC16F877 type microcontroller.

The following PDL describes the operation of the program:

**BEGIN**

Initialize the LCD

Initialize the microcontroller

**DO FOREVER**

Read RTD voltage and convert to digital

Calculate RTD resistance, $R_T$ using equation 4-17

Calculate temperature t using equation 4-18

Scale and format the temperature for the display

Display the temperature on LCD

Wait for 1 second

**ENDDO**

**END**

The complete program listing of the project is shown in Fig. 4.13. Variables *rtdv* and *rtdr* are the RTD voltage and resistance respectively. At the beginning of

```
/*********************************************************************
*
*
*     PROJECT:          PROJECT4
*     FILE:             PROJ4.C
*     DATE:             November 2001
*     PROCESSOR:        PIC16F877
*     COMPILER:         FED C
*
*
*This project measures the temperature with a Platinum RTD and
*then displays on a LCD every second in the format "nn.m C".
*
**********************************************************************/

#include <P16F877.h>
#include <displays.h>
#include <delays.h>
#include <strings.h>
#include <maths.h>

const int LCDPORT=&PORTB;                   //Define LCD port as Port B
void main()
{
          float rtdv,rtdr,temp,y,LSB;
          int msd,lsd;
          char temperature[6];
```

**Fig. 4.13** Program listing of the temperature measurement system

```
        LSB=5000.0/1024.0;                  //LSB in mV

        /* Initialize the LCD */
        LCD(-1);                            //Init. LCD to 1 line
        LCD(257);                           //Clear LCD and home

        while(1)                            //DO FOREVER
        {
             /* Configure the A/D */
             ADCON1 = 0x80;                 //Set 6 MSB to zero
             ADCON0 = 0x41;                 //Set A/D oscillator


             /* Start A/D conversion */
             ADCON0=0x45;                   //Start A/D
conversion
             while((ADCON0&4)!=0);          //Wait for conversion
             rtdv=256.0*(float)ADRESH+(float)ADRESL;   //Get RTD
voltage
             rtdv=rtdv*LSB/5.0;             //Scale RTD voltage
             rtdv=rtdv/1000.0;              //RTD in Volts

             /* Calculate RTD resistance rtdr (eqn 4-16) */
             rtdr = rtdv*1000.0/(5.0-rtdv);
             /* Calculate temperature temp (eqn 4-17) */

             y = 0.15274-(rtdr-100.0)*2310.0e-7;
             if(y>= 0)y = sqrt(y);

             temp = (y-0.39083)/(-0.0001155);
             /* Format for the display */
             LCD(257);                      //Clear LCD and home


             msd=temp;                                //msd digit
             lsd=10.0*(temp-msd);           //lsd digit
             iPrtString(temperature,msd);   //msd

             /* Insert ".", lsd digit, and "C" character */
             temperature[2]='.';
             cPrtString(temperature+3,lsd);
             temperature[4]='';
             temperature[5]='C';

             /* Display the temperature as "nn.m C" */
             LCDString(temperature);
             /* One second delay */
             Wait(1000);
        }                                                      //ENDDO

}
```

Fig. 4.13 (*Continued*)

the program the LCD is initialized and then the microcontroller A/D channel 0 is initialized. RTD voltage is read by the A/D converter and then scaled. RTD resistance *rtdr* is then calculated using equation 4-17. The measured temperature *temp* is calculated using equation 4-18. Notice that the *sqrt* library function is used to find the square-root of a floating point number. The decimal (*msd*) and the fractional (*lsd*) parts of the calculated temperature are then found, converted into string format and sent to the LCD display. This process is repeated forever with a one second delay between each output.

## 4.8 Exercises

1. The resistance of an RTD is 100 Ω at 0°C. If the A and B constants are as given below, calculate the temperature when the resistance is measured as 120 Ω.

$$A = 3.908 \times 10^{-1}$$

$$B = -5.775 \times 10^{-2}$$

$$R_0 = 100\,\Omega$$

$$R_t = 138\,\Omega$$

2. A platinum RTD is used in a temperature measurement system. If the A and B constants are as given in Exercise 1 above, calculate the resistance of this RTD when the temperature is 80°C.

3. The a, β, and d constants of a platinum RTD used to measure the temperature between −20°C and 80°C are specified by the manufacturers as:

$$a = 0.003850$$

$$\beta = 0.10863$$

$$d = 1.4999$$

   Calculate the standard constants A, B, and C.

4. Why is self-heating important in temperature sensor circuits?

5. Why is the lead resistance important in RTD sensor circuits? Explain how you can minimize the affects of this resistance.

6. What are the advantages of the thin film RTD sensors?

7. Modify the program code in Project 4.7 so that the temperature can be displayed both in °C and in °F.

8. Explain the differences between Class A and Class B standards, as applied to RTDs.

9. What is the temperature tolerance of a Class B standard RTD at +10°C? What will be the tolerance if the temperature is increased to +100°C.

10. Describe how the circuit in Project 4.7 can be improved so that the sensor can be located remotely from the microcontroller.

11. It is required to design a microcontroller based RTD temperature measurement system to measure the temperature of a liquid which varies between −15°C and 50°C. The microcontroller system can be located very close to the RTD sensor but the environment is electrically noisy with electric motors operating very close to the system. The temperature should be displayed on a LCD display every second with a resolution of 0.2°C. Design a microcontroller based temperature measurement system and show your complete circuit diagram. Write a program to read and display the temperature.

# Chapter 5

# Thermistor Temperature Sensors

## 5.1 Thermistor principles

The name thermistor derives from the words "thermal" and "resistor". Thermistors are temperature sensitive passive semiconductors which exhibit a large change in electrical resistance when subjected to a small change in body temperature. As shown in Fig. 5.1, thermistors are manufactured in a variety of sizes and shapes. Beads, discs, washers, wafers, and chips are the most widely used thermistor sensor types.

Disc thermistors are made by blending and compressing various metal oxide powders with suitable binders. The discs are formed by compressing under very

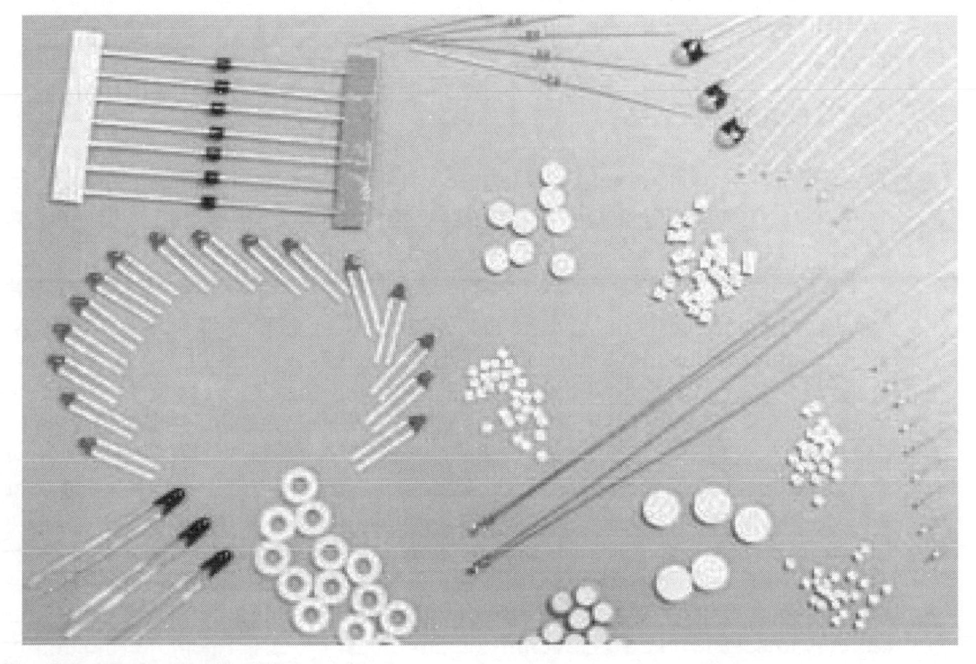

**Fig. 5.1** Some typical thermistors

high pressure on pelleting machines to produce round, flat ceramic bodies. An electrode material (usually silver) is then applied to the opposite sides of the disc to provide the contacts for attaching the lead wires. Sometimes a coating of glass or epoxy is applied to protect the devices from the environment and mechanical stresses. Finally, the thermistors are subjected to a special ageing process to ensure high stability of their values. Typical coated thermistors measure 2.0 mm to 4.0 mm in diameter.

Washer-shaped thermistors are a variation of the standard disc shaped thermistors, and they usually have a hole in the middle so that they can easily be connected to an assembly.

Bead thermistors have lead wires which are embedded in the ceramic material. They are manufactured by combining the metal oxide powders with suitable binders and then firing them in a furnace with the leads on. After firing, the ceramic body becomes denser around the wire leads. Finally, the leads are cut to create individual devices and a glass coating is applied to protect the devices from environmental effects and to provide long-term stability.

Chip thermistors are manufactured by using a technique similar to the manufacturing of ceramic chip capacitors. An oxide binder similar to the one used in making bead thermistors is poured into a fixture. The material is then allowed to dry into a flexible ceramic tape, which is then cut into smaller pieces and sintered at high temperatures into small wafers. These wafers are then diced into chips and the chips can either be used as surface mount devices, or leads can be attached for making discrete thermistors.

## 5.2 Thermistor types

Thermistors are generally available in two types: Negative Temperature Coefficient (NTC) thermistors, and Positive Temperature Coefficient (PTC) thermistors.

PTC thermistors are generally used in power circuits for in-rush current protection. Commercially there are two types of PTC thermistors. The first type consists of silicon resistors and are also known as “silistors”. These devices have a fairly uniform positive temperature coefficient during most of their operational ranges, but they also exhibit a negative temperature coefficient in higher temperatures. Silistors are usually used to temperature compensate silicon semiconductor devices. The other and more commonly used PTC thermistors are also known as “switching PTC” thermistors. These devices have a small negative temperature coefficient until the device reaches a critical temperature (also known as the “Curie” temperature). As this critical temperature is approached, the device shows a rising positive temperature coefficient of resistance as well as a large increase of resistance. These devices are usually used in switching applications to limit currents to safe levels. PTC thermistors are not used in temperature monitoring and control applications and thus will not be discussed further in this book.

NTC thermistors exhibit many desirable features for temperature measurement and control. Their electrical resistance decreases with increasing temperature (see Fig. 5.2) and the resistance-temperature relationship is very non-linear. Depending upon the type of material used and the method of fabrication, thermistors can be used within the temperature range of $-50°C$ to $+150°C$. The resistance of a thermistor is referenced to 25°C and for most applications the resistance at this temperature is between 100 Ω and 100 kΩ.

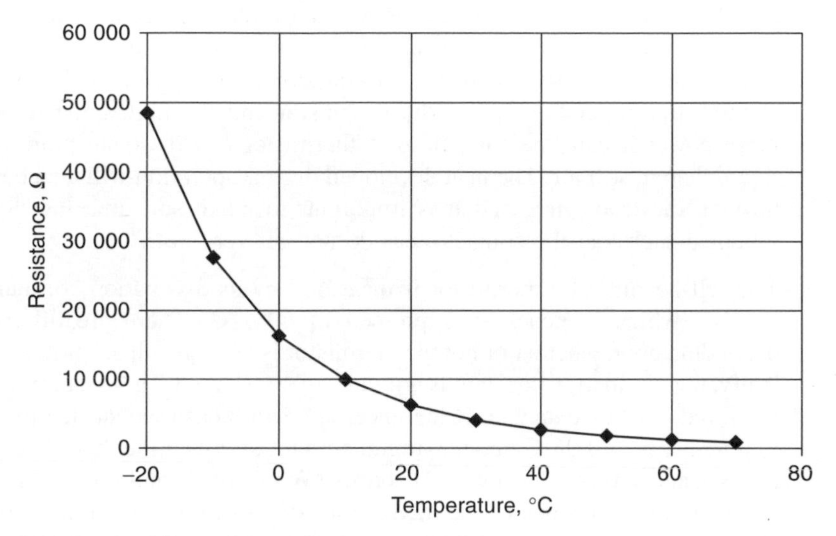


**Fig. 5.2** Typical thermistor R/T characteristics

The advantages of NTC thermistors are summarized below:

**Sensitivity**: One of the advantages of thermistors compared with thermocouples and RTDs is their relatively large change in resistance with temperature, typically −5% per °C.

**Small size**: Thermistors have very small sizes and this makes for a very rapid response to temperature changes. This feature is very important in temperature feedback control systems where a fast response is required.

**Ruggedness**: Most thermistors are rugged and can handle mechanical and thermal shock and vibration better than other types of temperature sensors.

**Remote measurement**: Thermistors can be used to sense the temperature of remote locations via long cables because the resistance of a long cable is insignificant compared to the relatively high resistance of a thermistor.

**Low cost**: Thermistors cost less than most of the other types of temperature sensors.

**Interchangeability**: Thermistors can be manufactured with very close tolerances. As a result of this it is possible to interchange thermistors without having to re-calibrate the measurement system.

## 5.3 Self-heating

A thermistor can experience self-heating as a result of the current passing through it. When a thermistor self-heats, the resistance reading becomes less than that of its true value and this causes errors in the measured temperature. The amount of self-heating is directly proportional to the power dissipated by the thermistor. In general, the larger the size of the thermistor and the higher the resistance, the more power it can dissipate. Also, a thermistor can dissipate more power in a liquid than in still air. The heat developed during operation will also be dissipated through the lead wires and it is important to make sure that the thermistor is mounted such that the contact areas do not get very hot.

The self-heating of a thermistor is measured by its dissipation constant (denoted by $d_{th}$), which is generally expressed in mW/°C. There are different values depending upon whether or not the thermistor is used in still air or in a liquid. Typically, the dissipation constant is around 1 mW/°C in still air and around 8 mW/°C in stirred oil. For example, if the uncertainty in measurement due to self-heat is required to be =0.05°C, the maximum amount of power the thermistor is allowed to dissipate is: 0.05°C × 1 mW/°C or 50 μW in still air and 400 μW in stirred oil. Knowing the resistance of the thermistor, we can calculate the maximum allowable current through the thermistor. Equation 5-1 shows the relationship between the dissipation constant and the thermal error due to self-heat of a thermistor:

$$P = T_e \cdot d_{th} \tag{5-1}$$

where P is the power dissipated in the thermistor, $T_e$ is the thermal error due to self-heat, and $d_{th}$ is the dissipation constant of the thermistor.

### Example 5.1

A thermistor is to be used as a sensor to measure the ambient temperature in a room. It is quoted that the dissipation constant in still air is 2 mW/°C. If the power dissipated by the thermistor is 200 μW, calculate the thermal error due to self-heat of the thermistor.

### Solution 5.1

The thermal error can be calculated by using equation 5-1.

From equation 5-1,

$$T_e = P/d_{th} \tag{5-2}$$

or,

$$T_e = 200 \times 10^{-6}/2 \times 10^{-3}$$

or,

$$T_e = 0.1^{\circ}C$$

# 5.4 Thermal time constant

Thermal time constant of a thermistor is the time required for a thermistor to reach 63.2% of a new temperature. Thermal time constant depends upon the type and mass of the thermistor and the method used for thermal coupling to its environment. A typical epoxy coated thermistor has a thermal time constant of 0.7 second in stirred oil and 10 seconds in still air.

# 5.5 Thermistor temperature-resistance relationship

Thermistor manufacturers usually provide the resistance-temperature characteristics of their devices either in the form of a table, a graph, or they provide some parameters which can be used to describe the behaviour of a thermistor. All different techniques are described in this section.

## 5.5.1 Temperature-resistance table

Some manufacturers provide tables which give the resistance of their devices at different temperatures. Table 5.1 is an example temperature-resistance table where the temperature is specified in increments of 10°C and the resistance at any temperature can be estimated by interpolation. Some manufacturers provide tables with temperature increments of only 1°C or even lower.

## 5.5.2 Steinhart–Hart equation

The *Steinhart–Hart equation* is an empirically developed polynomial which best represents the temperature-resistance relationships of NTC thermistors. The Steinhart–Hart equation is very accurate over the operating temperature of thermistors and it introduces errors of less than 0.1°C over a temperature range of −30°C to +125°C. To solve for temperature when resistance is known, the form of this equation is:

$$1/T_T = a + b\,Ln(R_T) + c[Ln(R_T)]^3 \qquad (5\text{-}3)$$

**Table 5.1** Typical thermistor temperature-resistance table

| Temperature (°C) | Resistance (Ω) |
|---|---|
| −30 | 88 500 |
| −20 | 48 535 |
| −10 | 27 665 |
| 0 | 16 325 |
| +10 | 9 950 |
| +20 | 6 245 |
| +30 | 4 028 |
| +40 | 2 663 |
| +50 | 1 801 |
| +60 | 1 244 |
| +70 | 876 |
| +80 | 627 |
| +90 | 457 |
| +100 | 339 |

where $T_T$ is the measured temperature (°K), $R_T$ is the resistance of the thermistor at temperature $T_T$, and a, b, and c are thermistor coefficients.

Equation 5-3 is usually written in the form:

$$T_T = \frac{1}{a + b\,Ln(R_T) + c[Ln(R_T)]^3} \tag{5-4}$$

The coefficients a, b, and c are sometimes given by the manufacturers. If these coefficients are not known, they can be calculated if the resistance is known at three different temperatures.

Using equation 5-3, and assuming that the resistances of the thermistor at temperatures $T_1$, $T_2$, and $T_3$ are $R_1$, $R_2$, and $R_3$ respectively, we can write:

$$1/T_1 = a + b\,Ln(R_1) + c[Ln(R_1)]^3 \tag{5-5}$$

$$1/T_2 = a + b\,Ln(R_2) + c[Ln(R_2)]^3 \tag{5-6}$$

$$1/T_3 = a + b\,Ln(R_3) + c[Ln(R_3)]^3 \tag{5-7}$$

Equations 5-5, 5-6, and 5-7 are three equations with three unknowns and these equations can be solved to find the thermistor coefficients a, b, and c.

The resistances at three different temperatures can either be measured or they can be obtained from the temperature-resistance tables as in Table 5.1. An example is given below to illustrate the method used.

**Example 5.2**

The temperature-resistance table of a thermistor is given in Table 5.1. Calculate the a, b, and c parameters of this thermistor and write down the Steinhart–Hart equation.

**Solution 5.2**

We can write three equations with three unknowns by taking three points on the resistance-temperature curve.

Using Table 5.1 and taking three different temperatures,

At 10°C $\quad T = 283°K \quad R_T = 9950\,\Omega$

At 20°C $\quad T = 303°K \quad R_T = 4028\,\Omega$

At 50°C $\quad T = 323°K \quad R_T = 1801\,\Omega$

now, using equations 5-5, 5-6, and 5-7 we can write:

$$1/283 = a + b\,\mathrm{Ln}(9950) + c[\mathrm{Ln}(9950)]^3 \tag{5-8}$$

$$1/303 = a + b\,\mathrm{Ln}(4028) + c[\mathrm{Ln}(4028)]^3 \tag{5-9}$$

$$1/323 = a + b\,\mathrm{Ln}(1801) + c[\mathrm{Ln}(1801)]^3 \tag{5-10}$$

or,

$$0.00353 = a + 9.2053b + 780.041c \tag{5-11}$$

$$0.00330 = a + 8.301b + 571.999c \tag{5-12}$$

$$0.00309 = a + 7.496b + 421.217c \tag{5-13}$$

Solving equations 5-11, 5-12, and 5-13, the coefficients are:

$$a = 4.34951 \times 10^{-3}$$

$$b = -2.89247 \times 10^{-4}$$

$$c = 2.36282 \times 10^{-6}$$

The Steinhart–Hart equation can now be written as:

$$T_T = \frac{10^6}{4349.51 - 289.247\ \mathrm{Ln}(R_T) + 2.36282[\mathrm{Ln}(R_T)]^3} \tag{5-14}$$

### 5.5.3 Using temperature-resistance characteristic formula

Some thermistor manufacturers give the resistances of their devices at 25°C and they also provide a thermistor temperature constant denoted by $\beta$. Equation 5-15 below can be used to calculate the resistance at any other temperature:

$$R_T = R_{25}\ \exp(\beta/T_T - \beta/T_{25}) \tag{5-15}$$

where $R_{25}$ is the resistance at 25°C, $T_{25}$ is the temperature (in °K) at 25°C, and $\beta$ is the temperature constant of the thermistor.

The temperature can be calculated using equation 5-15 as follows:

$$\frac{R_T}{R_{25}} = \exp\left(\frac{\beta}{T_T} - \frac{\beta}{T_{25}}\right) \tag{5-16}$$

or,

$$T_T = \frac{1}{1/\beta\ \mathrm{Ln}(R_T/R_{25}) + 1/T_{25}} \tag{5-17}$$

An example is given below to clarify the method used.

**Example 5.3**

The temperature constant of a thermistor is $\beta = 2910$. Also, the resistance of the thermistor at 25°C is $R_{25} = 1\,k\Omega$. This thermistor is used in an electrical circuit to measure the temperature and it is found that the resistance of the thermistor is 800 Ω. Calculate the temperature.

**Solution 5.3**

First of all, the temperature at 25°C is converted into °K

$$T_{25} = 25 + 273.15 = 298.15°K$$

We can now use equation 5-17.

From equation 5-17,

$$T_T = \frac{1}{1/2910\ \text{Ln}(800/1000) + 1/298.15} \quad (5\text{-}18)$$

or,

$$T_T = 310.24^\circ K$$

or,

$$T_T = 37.09^\circ C$$

**The β value**

The temperature constant β depends upon the material used in manufacturing the thermistor. This parameter also depends on temperature and manufacturing tolerances and as a result of this, equation 5-17 is only suitable for describing a restricted range around the rated temperature or resistance with sufficient accuracy. The β values for common NTC thermistors range from 2000k through 5000k. For practical applications, the more complicated Steinhart–Hart equation is more accurate.

Some manufacturers do not provide the value of β but they give the thermistor resistances at 25°C and also at 100°C. We can then calculate the value of β using equation 5-15:

$$\beta = \frac{T_T \cdot T_{25}}{T_T - T_{25}} \text{Ln}\frac{R_{25}}{R_T} \quad (5\text{-}19)$$

or,

$$\beta = \frac{(25 + 273.15)(100 + 273.15)}{75} \text{Ln}\,\frac{R_{25}}{R_{100}} \quad (5\text{-}20)$$

giving,

$$\beta = 1483.4\ \text{Ln}\frac{R_{25}}{R_{100}} \quad (5\text{-}21)$$

Thus, knowing the resistances at 25°C and also at 100°C we can use equation 5-21 to calculate the value of β.

**Example 5.4**

The manufacturer of a certain type of thermistor specifies the resistance of the device at 25°C and at 100°C as 1000 Ω and 600 Ω respectively. Calculate the temperature constant β of this thermistor.

**Solution 5.4**

The temperature constant β can be calculated using equation 5-21.

From equation 5-21,

$$\beta = 1483.4 \ \mathrm{Ln}\, 1.66$$

or,

$$\beta = 751.813$$

### 5.5.4 Thermistor linearization

Thermistor sensors are highly non-linear devices. It is possible to obtain very linear response from thermistors by connecting a resistor in parallel with the thermistor. The resistor value should be equal to the thermistor's resistance at the mid range temperature of interest. The combination of a thermistor and a parallel resistor has an S-shaped temperature-resistance curve with a turning point.

## 5.6 Practical thermistor circuits

In this section we shall be looking at various techniques of using NTC thermistors to measure temperature. Because of the very high temperature constant of the thermistor, accurate temperature measurements can be made with very simple electrical circuits. Typical applications include temperature control of ovens, freezers, rooms, temperature alarms, chemical process control and so on.

### 5.6.1 Constant current circuit

Figure 5.3 shows how a thermistor can be used in a very simple electrical circuit to measure temperature. A constant current I is passed through the thermistor which produces a voltage $V_T$ proportional to the current. By measuring this

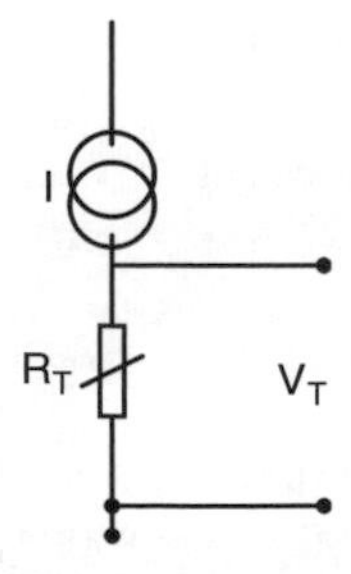


**Fig. 5.3** Constant current thermistor circuit

voltage and knowing the current, we can calculate the resistance of the thermistor as $R_T = V_T/I$ and hence the temperature can be calculated by using equations 5-3 or 5-15.

## 5.6.2 Constant voltage circuit

Figure 5.4 shows another simple thermistor circuit where a potential divider circuit is formed using a constant voltage source $V_S$. The voltage $V_T$ across the thermistor is calculated as:

$$V_T = V_S \frac{R_T}{R_S + R_T} \tag{5-22}$$

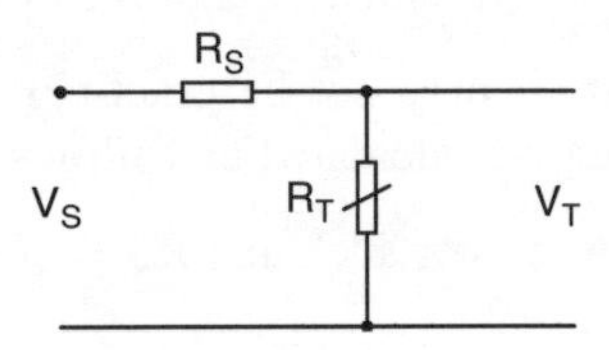


**Fig. 5.4** Constant voltage thermistor circuit

Using equation 5-22, the resistance of the thermistor can be found to be:

$$R_T = \frac{V_T \cdot R_S}{V_S - V_T} \tag{5-23}$$

Once the resistance is found, we can use either equation 5-3 or equation 5-15 to calculate the temperature.

## 5.6.3 Bridge circuit

Figure 5.5 shows how the temperature can be measured using a simple electrical bridge circuit. Bridge circuits have the advantage that under balance conditions the resistance of the arms are independent of the supply voltage. In this circuit,

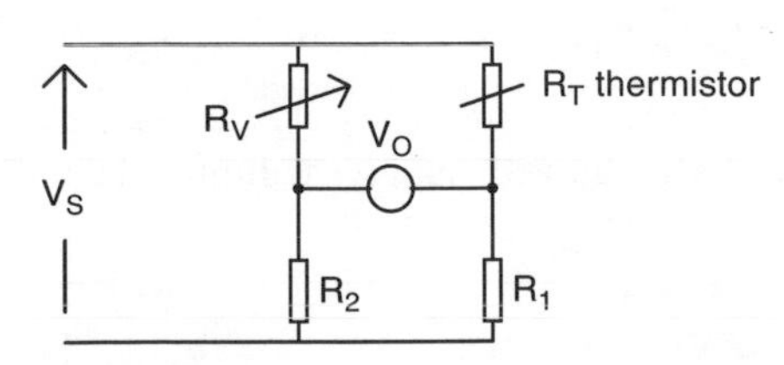


**Fig. 5.5** Thermistor bridge circuit

the thermistor is in one of the arms of the bridge and resistor $R_V$ is used to balance the bridge. In balance condition, the relationship between the resistors are:

$$\frac{R_T}{R_V} = \frac{R_1}{R_2} \tag{5-24}$$

and the thermistor resistance can be found as:

$$R_T = R_V \frac{R_1}{R_2} \tag{5-25}$$

The values of resistors $R_1$, $R_2$, and $R_V$ are all known and the temperature can be calculated by using $R_T$ in either equation 5-3 or equation 5-15.

In some applications it may not be practical to balance the bridge circuit. The value of $R_T$ can then be calculated as follows:

Let $R_V$ be a constant resistor such that $R_V = R_2$. When the bridge is not balanced, the value of $V_O$ is:

$$V_O = V_S \left( \frac{R_1}{R_1 + R_T} \right) - \frac{V_S}{2} \tag{5-26}$$

and the thermistor resistance is

$$R_T = R_1 \left( \frac{V_S - 2V_O}{V_S + 2V_O} \right) \tag{5-27}$$

### 5.6.4 Non-inverting operational amplifier circuit

Figure 5.6 shows how the thermistor can be used in a simple non-inverting operational amplifier circuit to measure temperature. In this circuit, the output voltage $V_O$ is found as:

$$V_O = V_S \frac{R_1}{R_1 + R_T} \tag{5-28}$$

where $V_S$ is the stabilized supply voltage. The thermistor resistance can be found as:

$$R_T = R_1 \left( \frac{V_S}{V_O} - 1 \right) \tag{5-29}$$

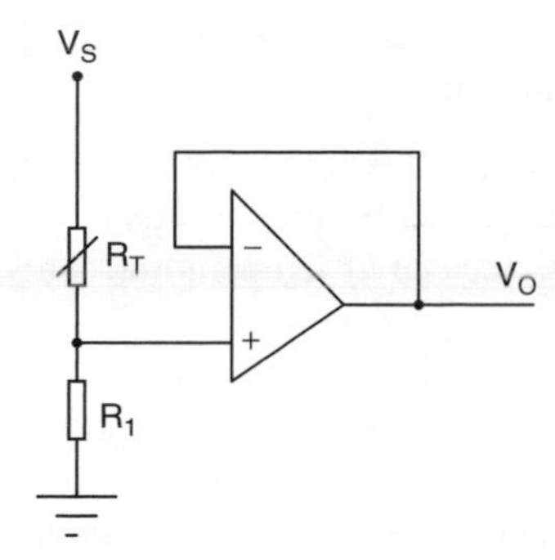


**Fig. 5.6** Non-inverting operational amplifier thermistor circuit

After finding $R_T$, we can use equation 5-3 or equation 5-15 to calculate the temperature.

### 5.6.5 Inverting operational amplifier circuit

Figure 5.7 shows how a thermistor can be connected in the feedback path of an inverting operational amplifier. In this circuit, the output voltage $V_O$ is:

$$V_O = V_S \frac{R_T}{R_S} \tag{5-30}$$

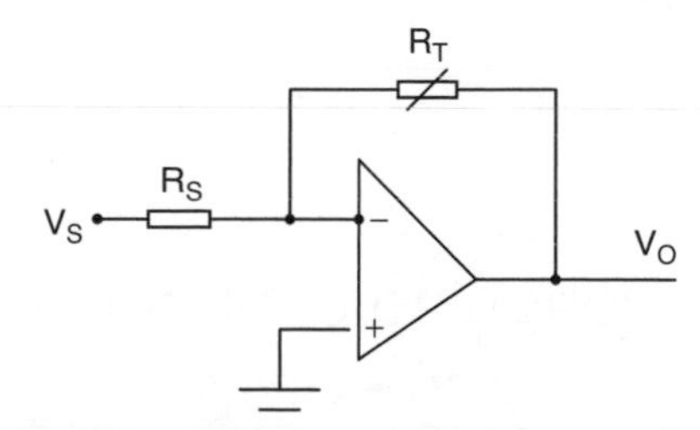


**Fig. 5.7** Inverting operational amplifier thermistor circuit

The thermistor resistance can be found as:

$$R_T = V_O \frac{R_S}{V_S} \tag{5-31}$$

## 5.7 Microcontroller based temperature measurement

Thermistors give analogue output voltages and the temperature can simply be displayed using an analogue meter (e.g. a voltmeter). Although this may be acceptable for some applications, accurate temperature measurement and display require digital techniques.

Figure 5.8 shows how the temperature can be measured using a microcontroller. The temperature is sensed by the thermistor and a voltage is produced which is proportional to the measured temperature. This voltage is filtered using a low-pass filter to remove any unwanted high frequency noise. The output of the filter is converted into digital form using a suitable A/D converter. The digitized voltage is then read by the microcontroller and the resistance of the thermistor is calculated. After this, the measured temperature can be calculated either by using a temperature-resistance table, or either equation 5-3 or equation 5-15. Finally, the temperature is displayed on a suitable digital display (e.g. an LCD). Although this approach is very simple and flexible, limitations arise in practice at both low and high temperatures. As the thermistor temperature decreases, its resistance increases and also the voltage across it increases. Practically the low temperature limit is reached when the voltage exceeds the maximum input voltage of the A/D converter. Similarly, when the thermistor temperature increases, its resistance decreases, so does the voltage across the thermistor. Practically the high temperature limit is reached when the voltage is less than the voltage resolution of the A/D converter. By varying the value of the constant current source in Fig. 5.3 or the fixed voltage source in Fig. 5.4, the useful temperature range of a thermistor can be extended for a given A/D converter.

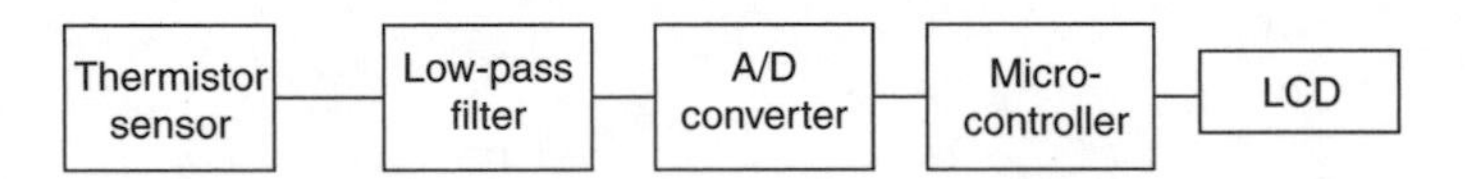


**Fig. 5.8** Microcontroller based temperature measurement

## 5.8 PROJECT – Designing a microcontroller based temperature measurement system using a thermistor

A complete temperature measurement system will be described in this section using a thermistor as the sensor. In this project the temperature is calculated using a microcontroller and then displayed every second on a LCD display. The complete circuit diagram and the program listing of the project are both provided in this section.

### 5.8.1 Specifications

The system is required to have the following specifications:

| | |
|---|---|
| Measuring temperature: | 0°C to +100°C with a few degrees accuracy |
| Self-heating error: | Less than 0.5°C |
| Display type: | LCD |

| | |
|---|---|
| Display format: | 6 characters displayed as "nn.m C" |
| Display update: | 1 second |
| Sensor: | Thermistor |
| System: | Microcontroller based |

## 5.8.2 Design

Many different types of thermistors can be selected to satisfy the above specifications. The thermistor selected for this design is the KED103BY, manufactured by Bowthorpe Thermistors. This is a miniature bead type thermistor having the following characteristics:

$$R_{25} = 10k$$

$$\beta = 4400$$

$$d_{th} = 9\,mW/^{\circ}C$$

We are assuming that a constant voltage source type design (Fig. 5.4) will be used to sense the temperature.

From the temperature-resistance graph (see Fig. 5.9) of this thermistor:

at $0^{\circ}C$ the resistance is $R_T = 40k$

at $100^{\circ}C$ the resistance is $R_T = 500\,\Omega$

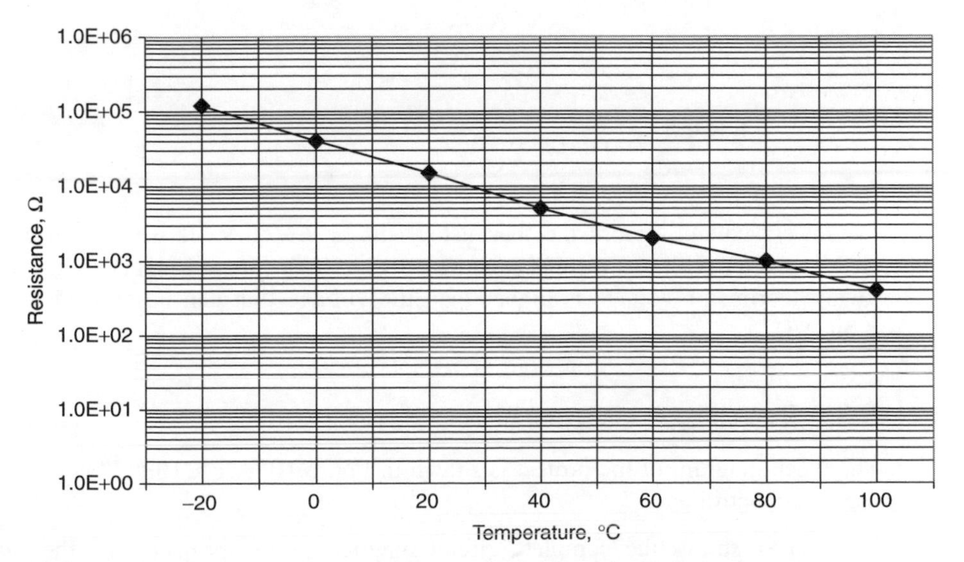


**Fig. 5.9** Temperature-resistance graph of the KED103BY thermistor

If we assume a maximum current of about 100 μA through the thermistor, the maximum power dissipation within the operating range will be:

$$P = I^2R \tag{5-32}$$

or,

$$P = (100 \times 10^{-6})^2 \times 40 \times 10^3 = 0.4\,\text{mW} \tag{5-33}$$

with a dissipation constant of 9 mW/°C, the maximum self-heating will be around $0.4/9 = 0.04°C$ which is well within the specified range.

At 0°C, the voltage across the thermistor is:

$$100 \times 10^{-6} \times 40 \times 10^3 = 4\,\text{V}$$

thus, the value of $R_S$ should be:

$$R_S = \frac{5 - 4}{100 \times 10^{-6}} = 10\text{k} \tag{5-34}$$

at 100°C, $R_T = 500\,\Omega$ and assuming a constant voltage source with a supply of 5 V, the voltage across the thermistor is:

$$V_T = V_S \frac{R_T}{R_S + R_T} \tag{5-35}$$

or,

$$V_T = \frac{5 \times 0.5}{0.5 + 40} = 0.0617 \tag{5-36}$$

the lowest voltage across the thermistor will be 61.7 mV. If we use an 8-bit A/D converter with a 5 V supply then 1 LSB = 5000/256 = 19.6 mV. Thus, the lowest thermistor voltage is higher than the voltage that can be applied to an 8-bit A/D.

### 5.8.3 The circuit diagram

The block diagram of the project is shown in Fig. 5.10. A PIC16F877 is used as the microcontroller.

Figure 5.11 shows the complete circuit diagram of the project. The thermistor is connected to a constant voltage source and the output is fed to channel

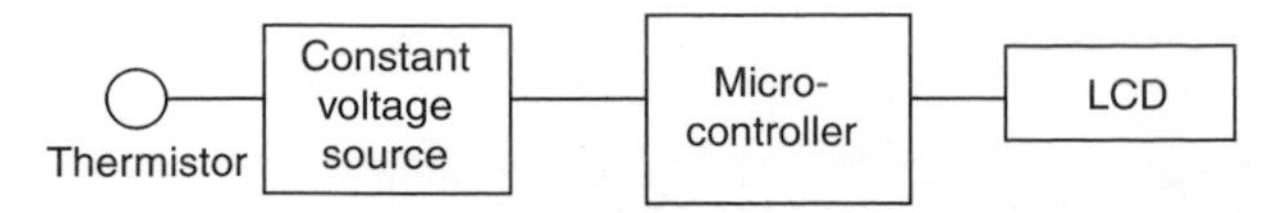


**Fig. 5.10** Block diagram of the project

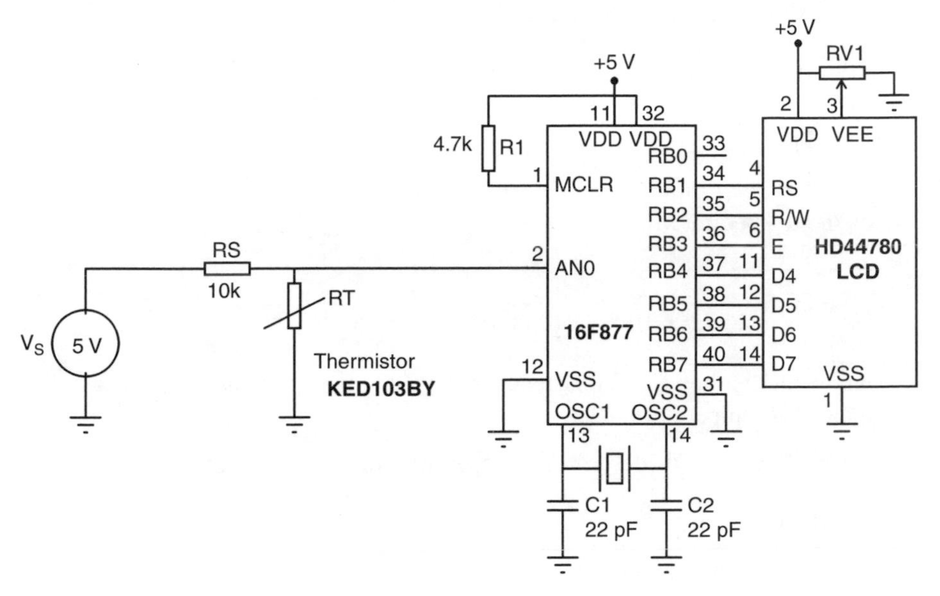


**Fig. 5.11** Circuit diagram of the project

AN0 of the PIC16F877 microcontroller. The LCD is connected to port B of the microcontroller where the measured temperature is displayed continuously in °C.

## 5.8.4 Operation of the circuit

The voltage across the thermistor is converted into digital form and the resistance of the thermistor ($R_T$) is calculated using equation 5-23. The measured temperature is then calculated using equation 5-17. Note that the Steinhart–Hart equation could also be used to calculate the temperature after finding $R_T$:

from equation 5-23,

$$R_T = \frac{V_T \cdot R_S}{V_S - V_T} \tag{5-23}$$

where $V_T$ is the voltage across the thermistor, $V_S = 5\,V$, and $R_S = 10k$.

Thus,

$$R_T = \frac{10^4 \cdot V_T}{5 - V_T}\ \Omega \tag{5-37}$$

from equation 5-17,

$$T_T = \frac{1}{1/\beta \ Ln(R_T/R_{25}) + 1/T_{25}} \qquad (5\text{-}17)$$

where $T_T$ is the measured temperature (°K), $R_T$ is the thermistor resistance (Ω), $\beta = 4400$, $T_{25} = 298.15$ (°K), and $R_{25} = 10k$.

Thus,

$$T_T = \frac{1}{1/4400 \ Ln(R_T/10\,000) + 1/298.15} \qquad (5\text{-}38)$$

or the temperature in °C is:

$$T_T = \frac{1}{2.272 \times 10^{-4} \ Ln(0.0001 R_T) + 3.354 \times 10^{-3}} - 273.15 \qquad (5\text{-}39)$$

Knowing the thermistor resistance $R_T$, we can use equation 5-39 to calculate the measured temperature.

The following PDL describes the operation of the program:

BEGIN

Initialize the LCD

Initialize the microcontroller

**DO FOREVER**

READ thermistor voltage and convert to digital

Calculate thermistor resistance $R_T$ using equation 5-37

Calculate temperature $T_T$ using equation 5-39

Display the temperature on LCD

Wait for 1 second

**ENDDO**

END

## 5.8.5 Program listing

The complete program listing of the project is given in Fig. 5.12. Floating point variables *tv* and *tr* are the thermistor voltage and the resistance respectively. After initializing the LCD and the microcontroller, an endless loop is formed and the A/D conversion is started. The A/D converter is 10-bits wide

```
/*****************************************************************
*
*
*     PROJECT:     PROJECT5
*     FILE:        PROJ5.C
*     DATE:        November 2001
*     PROCESSOR:   PIC16F877
*     COMPILER:    FED C
*
*
* This project measures the temperature with a Thermistor and
* then displays on a LCD every second in the format "nn.m C".
*
******************************************************************/

#include <P16F877.h>
#include <displays.h>
#include <delays.h>
#include <strings.h>
#include <maths.h>

const int LCDPORT=&PORTB;                    //Define LCD port as Port B
const float LSB = 5000/1024;                 //LSB in mV


void main()
{
            float tv,tr,temp,y;
            int msd,lsd;
            char temperature[6];

            /* Initialize the LCD */
            LCD(-1);                         //Init. LCD to 1 line
            LCD(257);                        //Clear LCD and home

            /* Initialize the microcontroller */
            ADCON1 = 0x80;                   //Set 6 MSB bits to
zero
            ADCON0 = 0;                      //Select AN0 input
            ADCON0 = 0x41;                   //SET A/D oscillator
            while(1)                         //DO FOREVER
            {
                  /* Start A/D conversion */
                  ADCON0=0x45;               //Start A/D
conversion
                  while(ADCON0 & 1 == 0);  //Wait for conversion
                  tv = 256*ADRESH + ADRESL;//Get thermistor voltage
                  tv=tv*LSB;                 //Scale thermistor
voltage

                  /* Calculate thermistor resistance tr (eqn 5-37)*/
                  tr = 5.0*tv*10000.0/(5.0-tv);

                  /* Calculate temperature temp (eqn 5-39)*/
                  y = log(0.0001*tr);
```

**Fig. 5.12** Program listing of the temperature measurement system

```
                y = 3.354*1e-3+y*2.272*1e-4;
                temp = 1.0/y;

                /* Format for the display */
                LCD(257);                       //Clear LCD and home

                msd=temp;                                    //msd digit
                lsd=10.0*(temp-lsd);            //lsd digit

                iPrtString(temperature,msd);    //msd

                /* Insert ".", lsd digit, and "C" character */
                temperature[2]='.';
                cPrtString(temperature+3,lsd);
                temperature[4]=' ';
                temperature[5]='C';

                /* Display the temperature as "nn.m C" */
                LCDString(temperature);

                /* One second delay */
                Wait(1000);
            }                                                //ENDDO
}
```

**Fig. 5.12** *(Continued)*

and the upper 2-bits of the conversion are stored in PIC register *ADRESH*. The lower 8-bits are stored in register *ADRESL*. The thermistor voltage *tv* is calculated by multiplying ADRESH by 256 and then adding *ADRESL*. The voltage is then converted to millivolts by multiplying with the scale factor of 5000/1024 (full scale 10-bits is 5000 mV). The thermistor resistance *tr* is then calculated using equation 5-37. Finally, the measured temperature is calculated using equation 5-39 and stored in floating point variable *temp*. This variable is configured as two decimal digits and a digit after the decimal point and is converted into a string called *temperature*. The characters "C" are added to the end of this string and the resultant temperature is displayed on the LCD every second.

## 5.9 Exercises

1. A thermistor is to be used as a sensor to measure the ambient temperature in a room. It is quoted that the dissipation constant in still air is 1 mW/°C. If the power dissipated by the thermistor is 100 μW, calculate the thermal error due to self-heat of the thermistor.

2. It is required to design a thermistor based temperature measurement system. The temperature constant of the thermistor used is 2 mW/°C. If the thermal

error due to self-heating is not to exceed 0.5°C, calculate the maximum power that can be dissipated by the thermistor.

3. Explain the term "thermal time constant".

4. The temperature-resistance table of a thermistor is given in Table 5.1. Calculate the a, b, and c parameters of this thermistor using the temperature values 20°C, 30°C, and 40°C. Write down the Steinhart–Hart equation.

5. Calculate the resistance of the thermistor at 50°C using the equation derived in exercise 4.

6. The temperature constant of a thermistor is $\beta = 3000$. Also, the resistance of the thermistor at 25°C is $R_{25} = 1.1\,k\Omega$. This thermistor is used in an electrical circuit to measure the temperature and it is found that the resistance of the thermistor is 600 Ω. Calculate the temperature.

7. The manufacturer of a certain type of thermistor specifies the resistance of the device at 25°C and at 100°C as 1200 Ω and 500 Ω respectively. Calculate the temperature constant β of this thermistor.

8. Draw the circuit diagram of a thermistor temperature measurement system using an operational amplifier. What are the advantages of using an operational amplifier?

9. Modify the temperature measurement Project in 5.8 so that the temperature can be displayed both in °C and in °F.

10. Modify the temperature measurement Project in 5.8 so that the temperature can be calculated using a temperature-resistance table.

11. Modify the temperature measurement Project in 5.8 so that the temperature can be calculated using the Steinhart–Hart equation.

# Chapter 6

# Integrated Circuit Temperature Sensors

Integrated circuit temperature sensors are semiconductor devices fabricated in a similar way to other semiconductor devices such as microcontrollers. There are no generic types like the thermocouples or RTDs but some popular devices are manufactured by more than one manufacturer.

Integrated circuit temperature sensors differ from other sensors in some fundamental ways:

- These sensors have relatively small physical sizes.
- Their outputs are linear (e.g. 10 mV/°C).
- The temperature range is limited (e.g. −40°C to +150°C).
- The cost is relatively low.
- These sensors can include advanced features such as thermostat functions, built-in A/D converters and so on.
- Often these sensors do not have good thermal contacts with the outside world and as a result it is usually more difficult to use them other than in measuring the air temperature.
- A power supply is required to operate these sensors.

Integrated circuit semiconductor temperature sensors can be divided into the following categories:

A. Analog temperature sensors

B. Digital temperature sensors

Analog sensors are further divided into:

a. Voltage output temperature sensors

b. Current output temperature sensors

Analog sensors can either be directly connected to measuring devices (such as voltmeters), or A/D converters can be used to digitize the outputs so that they can be used in computer based applications.

Digital temperature sensors usually provide $I^2C$ bus, SPI bus, or some other 3-wire interface to the outside world.

## 6.1 Voltage output temperature sensors

These sensors provide a voltage output signal which is proportional to the temperature measured. There are many types of voltage output sensors and Table 6.1 gives some popular sensors. LM35 is a 3-pin device and it has two versions and they both provide a linear output voltage of 10 mV/°C. The temperature range of the "CZ" version is −20°C to +120°C while the "DZ" version only covers the range 0°C to +100°. The accuracy of LM35 is around ±1.5°.

**Table 6.1** Popular voltage output temperature sensors

| Sensor | Manufacturer | Output | Maximum error | Temperature range |
|---|---|---|---|---|
| LM35 | National Semiconductors | 10 mV/°C | ±1°C | −20°C to +120°C |
| LM34 | National Semiconductors | 10 mV/°F | ±3°F | −20°C to +120°C |
| LM50 | National Semiconductors | 10 mV/°C<br>500 mV offset | ±3° | −40°C to +125°C |
| LM60 | National Semiconductors | 6.25 mV/°C<br>424 mV offset | ±3°C | −40°C to +125°C |
| S-8110 | Seiko Instruments | −8.5 mV/°C | ±2.5°C | −40°C to +100°C |
| TMP37 | Analog Devices | 20 mV/°C | ±3°C | +5°C to +100°C |

LM34 is similar to LM35, but its output is calibrated in degrees Fahrenheit as 10 mV/°F.

LM50 can measure negative temperatures without any external components. The LM50's output voltage has a 10 mV/°C slope, and a 500 mV offset, i.e. the output voltage $V_O$ is:

$$V_O = 10\,\mathrm{mV}/^\circ\mathrm{C} + 500\,\mathrm{mV} \tag{6-1}$$

Thus, the output voltage is 500 mV at 0°C, 100 mV at −40°, and 1.5 V at +100°C.

LM60 gives a linear output of 6.25 mV/°C and it can operate with a supply voltage as low as 2.7 V. The temperature range of this sensor is −40°C to +125°C.

S-8110 has a negative temperature coefficient and it gives −8.5 mV/°C. This sensor can operate at very low currents (10 μA) and the temperature range is −40°C to +100°C.

Analog Devices TMP37 is a high sensitivity sensor with a linear output voltage of 20 mV/°C and an operating temperature range of +5°C to +100°C.

### 6.1.1 Applications of voltage output temperature sensors

The popular LM35DZ temperature sensor is taken as an example in this section. As shown in Fig. 6.1, this is a 3-pin sensor. The maximum supply voltage is +35 V, but the sensor is normally operated at +5 V. When operated at +5 V, the supply current is around 80 μA. The typical accuracy is ±0.6°C at +25°C.

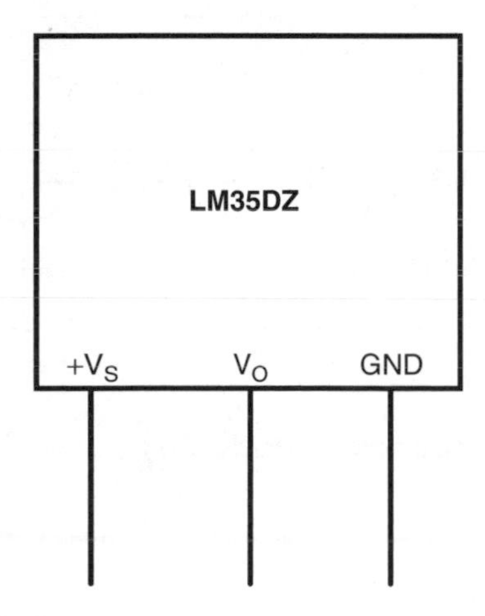


**Fig. 6.1** LM35DZ is a 3-pin sensor

Figure 6.2 shows how the LM35DZ can be used to measure the temperature. The device is simply connected to a 4 V to 20 V power supply and the output voltage is a direct indication of the temperature in 10 mV/°C. A simple voltmeter can be connected and calibrated to measure the temperature directly. Alternatively, an A/D converter can be used to digitize the output voltage so that the sensor can be used in computer based applications.

Care should be taken when driving capacitive loads, such as long cables. To remove the effect of such loads, the circuit shown in Fig. 6.3 should be used.

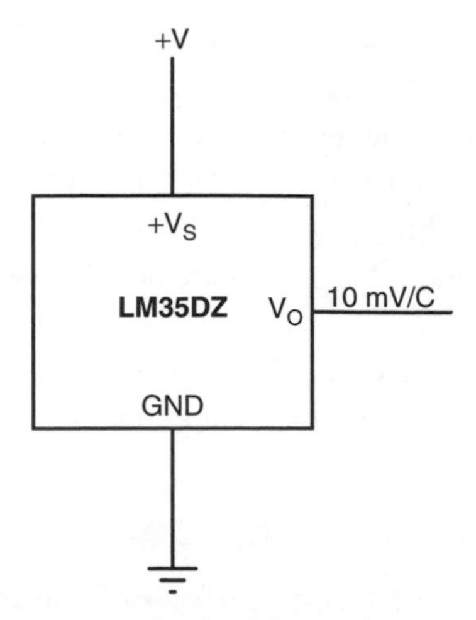


**Fig. 6.2** Using the LM35DZ to measure temperature

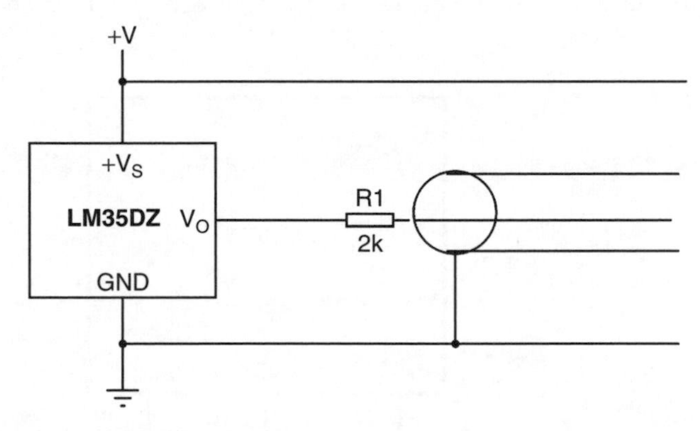


**Fig. 6.3** Driving capacitive loads

Notice that a resistor is added to the output of the sensor, making this circuit suitable for connection to high impedance loads only.

## 6.2 Current output temperature sensors

Current output sensors act as high impedance, constant current sources, giving an output current which is proportional to the temperature. The devices operate typically between 4 V and 30 V and produce an output current of $1\,\mu A/^\circ K$. Table 6.2 gives a list of some popular current output sensors.

AD590 has an operating temperature range of $-55^\circ C$ to $+150^\circ C$ and produces an output current of $1\,\mu A/^\circ K$.

**Table 6.2** Popular current output temperature sensors

| Sensor | Manufacturer | Output | Maximum error | Temperature range |
|---|---|---|---|---|
| AD590 | Analog Devices | 1 μA/°K | ±5.5°C | −55°C to +150°C |
| AD592 | Analog Devices | 1 μA/°K | ±1°C | −25°C to +105°C |
| LM134 | National Semiconductors | 0.1 μA/°K to 4 μA/°K | ±3°C | −25°C to +100°C |

AD592 is a more accurate sensor with a 1 μA/°K and the temperature range is −25°C to +105°C. The maximum error within the operating range is ±1°C.

LM134 is a programmable sensor with an output current 0.1 μA/°K to 4 μA/°K. The sensitivity is set using a single external resistor. The temperature range of this sensor is −25°C to +100°C. LM134 typically needs only 1.2 V supply voltage, so it can be used in portable applications where the power is usually limited.

## 6.2.1 Applications of current output temperature sensors

The popular AD590 is taken as an example here. This device gives an output current which is directly proportional to the temperature, i.e.

$$I = k^{*}T \tag{6-2}$$

where I is in μA, T in Kelvin, and k is the constant of proportionality (k = μA/°K).

Figure 6.4 shows a typical application of the AD590. A constant voltage source is used to supply the circuit. The voltage across the resistor is measured and then the current is calculated. The temperature can then be calculated using equation 6-2:

$$I_S = \frac{V_R}{R} \tag{6-3}$$

Using equation 6-2:

$$T = \frac{V_R}{kR}$$

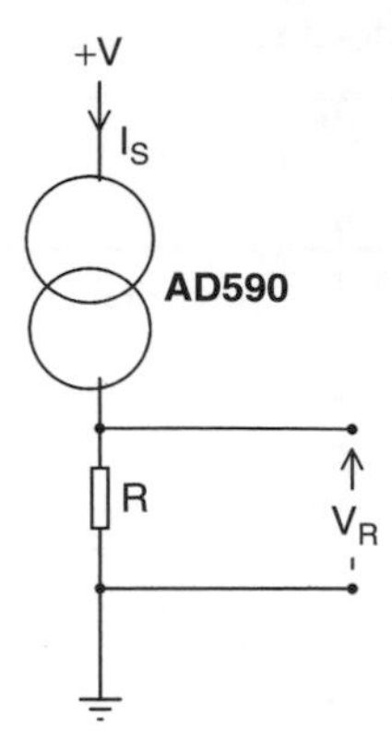


Fig. 6.4 Using AD590 to measure temperature

Assuming k = 1 μA/°K,

$$T = \frac{V_R}{R} \times 10^6 - 273.15 \tag{6-4}$$

where T is in °C, R is in ohms, and $V_R$ is in volts.

## 6.3 Digital output temperature sensors

The digital output temperature sensors produce digital outputs which can be interfaced directly to computer-based equipment. The outputs are usually non-standard and the temperature can be extracted by using suitable algorithms. Table 6.3 gives a list of some popular digital output temperature sensors.

LM75 operates within the temperature range of −55°C to +125°C and provides an $I^2C$ bus compatible 9-bit serial output (1 sign bit and 8 magnitude bits), resulting in 0.5°C resolution. The device can be programmed to monitor the temperature. It can also set an output pin high or low if the temperature exceeds a pre-programmed value. The temperature can be read by $I^2C$ bus compatible computer equipment. The $I^2C$ bus algorithm can also be developed on microcontrollers. For example, the FED C compiler has built-in functions which support the $I^2C$ bus interface. LM75 has an addressable, multi-drop connection feature which enables a number of similar sensors to be connected over the bus.

TMP03 provides PWM (Pulse Width Modulation) output and the computer system it is connected to measures the width of the pulse to find the temperature. The temperature range is −25°C to +100°C.

DS1620 is a temperature sensor which also incorporates digitally programmable thermostat outputs. The device provides 9-bit temperature readings which indicate

**Table 6.3** Popular digital output temperature sensors

| Sensor | Manufacturer | Output | Maximum error | Temperature range |
|---|---|---|---|---|
| LM75 | National Semiconductors | $I^2C$ | ±3°C | −55°C to +125°C |
| TMP03 | Analog Devices | PWM | ±4°C | −25°C to +100°C |
| DS1620 | Dallas Semiconductors | 2 or 3-wire | ±0.5°C | −55°C to +125°C |
| AD7814 | Analog Devices | SPI | ±2°C | −55°C to +125°C |
| MAX6575 | Maxim | Single wire | ±0.8°C | −40°C to +125°C |

the temperature of the device. Temperature settings and temperature readings are all communicated to/from the DS1620 over a 2 or 3-wire interface.

AD7814 operates within the temperature range of −55°C to +125°C and provides SPI bus compatible 10-bit resolution output. The temperature can be read by SPI bus compatible computer equipment.

MAX6575 sensor enables multiplexing of up to eight sensors on a simple single-wire bi-directional bus. Temperature is sensed by measuring the time delays between the falling edge of an external triggering pulse and the falling edge of the subsequent pulse delays reported from the devices. Different sensors on the same line use different timeout multipliers to avoid overlapping signals. The temperature range of MAX6575 is −40°C to +125°C.

## 6.3.1 Applications of digital output temperature sensors

The popular DS1620 temperature sensor is considered in this section. Figure 6.5 shows the sensor pin configuration. DS1620 is a digital thermometer and thermostat IC that provides 9-bits of serial data to indicate the temperature of the device. VDD is the power supply which is normally connected to a +5 V supply. DQ is the data input/output pin. CLK is the clock input. RST is the reset input. The device can also act as a thermostat. THIGH is driven logic high if the DS1620's temperature is greater than or equal to a user defined temperature TH. Similarly, TLOW is driven logic high if the DS1620's temperature is less than or equal to a user defined temperature TL. TCOM is driven high when the temperature exceeds TH and stays high until the temperature falls below TL. User defined temperatures TL and TH are stored in a non-volatile memory of the device so that they are not lost even after removal of power.

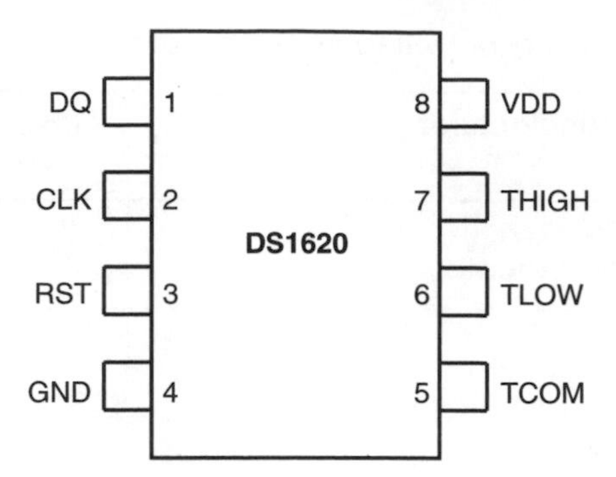


**Fig. 6.5** Pin configuration of DS1620

Data is output from the device as 9-bits, with the LSB sent out first. The temperature is provided in 2's complement format from −55°C to +125°C, in steps of 0.5°C. Table 6.4 shows the relationship between the temperature and data output by the device.

**Table 6.4** Temperature/data relationship of DS1620

| Temp. (°C) | Digital output (binary) | Digital output (hex) | 2's complement | Digital output (decimal) |
|---|---|---|---|---|
| +125 | 0 11111010 | 0FA | - | 250 |
| +25 | 0 00110010 | 032 | - | 50 |
| 0.5 | 0 00000001 | 001 | - | 1 |
| 0 | 0 00000000 | 000 | - | 0 |
| −0.5 | 1 11111111 | 1FF | 001 | 511 |
| −25 | 1 11001110 | 1CE | 032 | 462 |
| −55 | 1 10010010 | 192 | 06E | 402 |

Data input and output is through the DQ pin. When the RST input is high, serial data can be written or read by pulsing the clock input. Data is written or read from the device in two parts. First, a protocol is sent and then the required data is read or written. The protocol is 8-bit data and the protocol definitions are given in Table 6.5. For example, to write the thermostat value TH, the hexadecimal protocol data 01 is first sent to the device. After issuing this command, the next nine clock pulses clock in the 9-bit temperature limit which will set the threshold for operation of the THIGH output.

For example, the following data (in hexadecimal) should be sent to the device to set it for a TH limit of +50°C and TL limit of +20°C and then subsequently to start conversion:

**Table 6.5** DS1620 Protocol definitions

| Protocol | Protocol data (hex) |
|---|---|
| Write TH | 01 |
| Write TL | 02 |
| Write configuration | 0C |
| Stop conversion | 22 |
| Read TH | A1 |
| Read TL | A2 |
| Read temperature | AA |
| Read configuration | AC |
| Start conversion | EE |

01 Send TH protocol

64 Send TH limit 50 (64 hex = 100 decimal)

02 Send TL protocol

28 Send TL limit of 20 (28 hex = 40 decimal)

EE Send conversion start protocol

A configuration/status register is used to program various operating modes of the device. This register is written with protocol 0C (hex) and the status is read with protocol AC (hex). Some of the important configuration/status register bits are as follows:

Bit 0: This is the 1 shot mode. If this bit is set, DS1620 will perform one temperature conversion when the start protocol is sent. If this bit is 0, the device will perform continuous temperature conversions.

Bit 1: This bit should be set to 1 for operation with a microconroller or microprocessor.

Bit 5: This is the TLF flag and is set to 1 when the temperature is less than or equal to TL.

Bit 6: This is the THF flag and is set to 1 when the temperature is greater then or equal to TH.

Bit 7: This is the DONE bit and is set to 1 when a conversion is complete.

## 6.4 PROJECT – Using a digital output sensor to measure the temperature

In this section we shall be looking at the design of a complete temperature measurement system using a digital output sensor. We shall be using the popular DS1620 described in Section 6.3.1.

### 6.4.1 The hardware

Assume that the requirement is to measure the temperature within the range 0°C to 99°C with ±1°C accuracy. Also assume that the temperature is to be displayed on a LCD display in the format "nn C" (e.g. 25°C).

Figure 6.6 shows the block diagram of the design. The complete circuit diagram is shown in Fig. 6.7. A low-cost PIC16F84 type microcontroller is used in the design. The RST and the CLK inputs of the DS1620 are connected to port pins RA0 and RA1 respectively. DQ input/output of DS1620 is connected to the RA2 pin of the microcontroller. The LCD display is connected to Port B of the microcontroller.

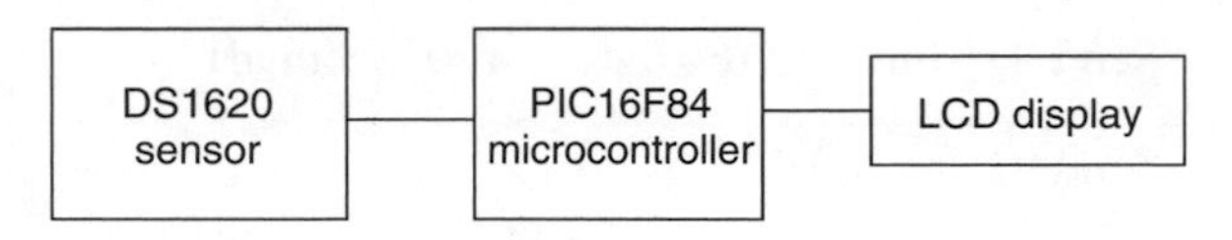


**Fig. 6.6** Block diagram of the project

### 6.4.2 The software

The functions of the program are summarized in the following PDL:

```
BEGIN
        Initialize the LCD
        Initialize the Microcontroller
        Configure the DS1620
        DO FOREVER
                Read temperature
                Format the data for the display
                Send data to the LCD
                Wait one second
        ENDDO
END
```

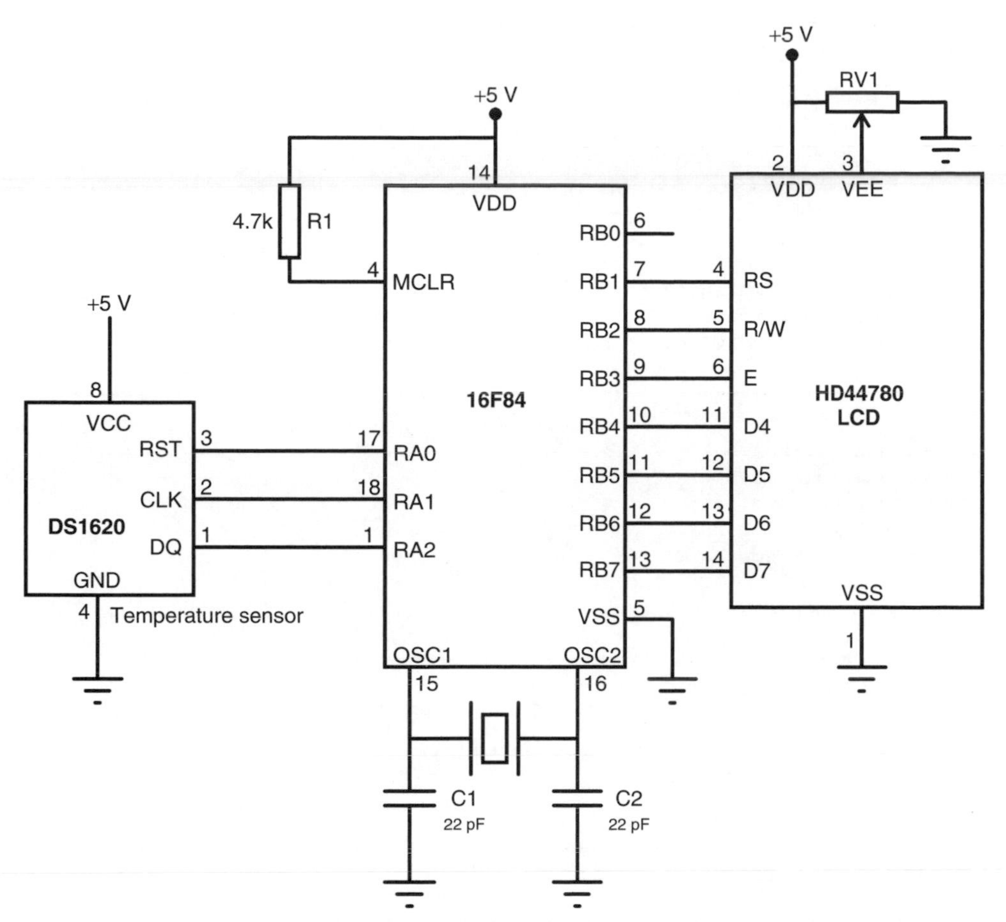


**Fig. 6.7** Circuit diagram of the project

The software listing is shown in Fig. 6.8. DS1620 pins *RST*, *CLK*, and *DQ* are assigned to Port A bits 0, 1, and 2 respectively. DS1620 functions are then assigned to program variables. The main program initializes the LCD display and defines the I/O configuration of Port A. Procedure *configure_ds1620* is then called to configure the DS1620 for continuous operation with a microcontroller. Procedure *start_conversion* sends protocol number *EE* (hexadecimal) to the chip to start the temperature conversion process. Inside the *while* loop the temperature is read by calling procedure *read_temperature*. This procedure sends the read_temperature protocol to the DS1620 and then reads the temperature by clocking in the serial data. The temperature is stored in variable *tmp*, which is converted into a character string called *temperature* using the built-in library function *iPrtString*. The characters "C" are then added to this string and the string is sent to the LCD display every second, using the built-in library function *LCDString*.

```
/********************************************************************
*
*
*     PROJECT:     PROJECT6
*     FILE:        PROJ6.C
*     DATE:        November 2001
*     PROCESSOR:   PIC16F84
*     COMPILER:    FED C
*
*
* This project measures the temperature with a DS1620 type
* integrated circuit (IC) sensor.  The temperature is then
* displayed on a LCD every second in the format "nn C".
*
********************************************************************/

#include <P16F84.h>
#include <displays.h>
#include <delays.h>
#include <datalib.h>
#include <strings.h>

const int LCDPORT=&PORTB;                        //Define LCD port as Port B

/* Port Defines */
#define RST PA.B0
#define CLK PA.B1
#define DQ  PA.B2

/* DS1620 Functions */
unsigned char write_config = 0x0c;
unsigned char start_conv   = 0xee;
unsigned char read_temp    = 0xaa;

void write_ds1620_bit(unsigned char b)
/* This function sends a data bit to DS1620 thermometer IC */
{
      DQ = b;
      CLK = 0;
      CLK = 1;
      DQ = 1;
}


unsigned char read_ds1620_bit()
/* this function reads a data bit from DS1620 */
{
      unsigned char b;

      CLK = 0;
      b = DQ;
      CLK = 1;
      return (b);
}
```

**Fig. 6.8** Program listing of the project

```
void write_to_ds1620(unsigned char ds1620_function,
      unsigned char ds1620_data,
      unsigned char bit_count)
/* This function writes data/configuration to DS1620 */
{
      unsigned char i,this_bit;

      RST = 1;

      /* Send function ... */
      for(i=0;i<8;i++)
      {
            this_bit = ds1620_function >> i;
            this_bit = this_bit & 0x01;
            write_ds1620_bit(this_bit);
      }

      /* Send data ... */
      for(i=0;i<bit_count;i++)
      {
            this_bit = ds1620_data >> i;
            this_bit = this_bit & 0x01;
            write_ds1620_bit(this_bit);
      }
      RST = 0;
}


unsigned int read_from_ds1620(unsigned char ds1620_function,
      unsigned char bit_count)
/* This function reads data/configuration from DS1620 */
{
      unsigned char i,this_bit;
      unsigned int ds1620_data;

      ds1620_data = 0;
      RST = 1;

      /* Send function */
      for(i=0;i<8;i++)
      {
            this_bit = ds1620_function >> i;
            this_bit = this_bit & 0x01;
            write_ds1620_bit(this_bit);
      }

      /* Read data */
      for(i=0;i<bit_count;i++)
      {
            ds1620_data = ds1620_data | read_ds1620_bit() << i;
      }
      RST = 0;
      return(ds1620_data);
}
```

**Fig. 6.8** (*Continued*)

```
void main()
{
            char temperature[4];
            unsigned int temp;

            /* Initialize the LCD */
            LCD(-1);                        //Init. LCD to 1 line
            LCD(257);                       //Clear LCD and home

            /* Initialize the microcontroller */
            TRISA = 0;

            /* Configure DS1620 for continuous operation */
            write_to_ds1620(write_config,2,8);

            /* Start conversion */
            write_to_ds1620(start_conv,0,0);

            while(1)                                      //DO FOREVER
            {
                  temp = read_from_ds1620(read_temp,9);   //in ds_1620_temp
                                            //only
positive temp
                  temp = temp /2;           //extract
actual temp
                  /* Format for the display */
                  LCD(257);                 //Clear LCD and
home
                  iPrtString(temperature,temp);

                  /* Insert ".", lsd digit, and "C" character */
                  temperature[2]=' ';
                  temperature[3]='C';

                  /* Display the temperature as "nn C" */

                  LCDString(temperature);

                  /* One second delay */
                  Wait(1000);
            }                                                        //ENDDO

}
```

**Fig. 6.8** *(Continued)*

## 6.5 Exercises

1. What are the advantages and disadvantages of integrated circuit temperature sensors?
2. Draw the block diagram of a microcontroller based temperature measurement system using a LM35 type voltage output temperature sensor.

3. Show how you can measure the temperature using a current output temperature sensor and a microcontroller.

4. You are required to design a low-cost microcontroller based temperature measurement system using the minimum number of I/O pins. The temperature is to be displayed on a LCD display. Draw the circuit diagram of such a system.

5. You are required to design a microcontroller based temperature measurement system to monitor the temperature of 8 ovens. The output should be updated and displayed using a single LCD display. Draw the circuit diagram of such a system, using a digital output temperature sensor.

6. Show how you can modify exercise 1 to monitor the temperature of up to 4 ovens. Give the circuit diagram and the program code of your design.

7. Modify exercise 2 so that the temperature data is output in serial form from an output pin of the microcontroller.

8. Modify exercise 2 to store the maximum and the minimum values of the temperature in the EEPROM of the PIC16F84 microcontroller.

# Chapter 7

# Digital Control Systems and the z-transform

The use of a digital computer as a controller (compensator) device has grown rapidly during the last decade. One of the main reasons for this growth is the low price and the high reliability of digital computers. The total number of computer control systems installed in industry has grown in the last few decades. Currently there are approximately 100 million control systems around the world using computers. If we consider only the complex computer control systems, such as aircraft control, the number of computer control systems is around 20 million.

Digital control systems are used in many applications, including chemical processes, traffic control, aircraft control, radar control, machine tools and so on. The advantages of using digital control include improved accuracy and programmability. The control algorithm can easily be configured in software.

Figure 7.1 shows the block diagram of a single-loop digital control system. As shown in the figure, the reference input is received in digital form and error signal is also in digital form. The computer performs calculations and produces an output control signal which is also in digital form. A digital to analog (D/A) converter is used to convert the computer output into analog form in order to drive the process. The computer is programmed to provide an output so that the process is at the desired performance.

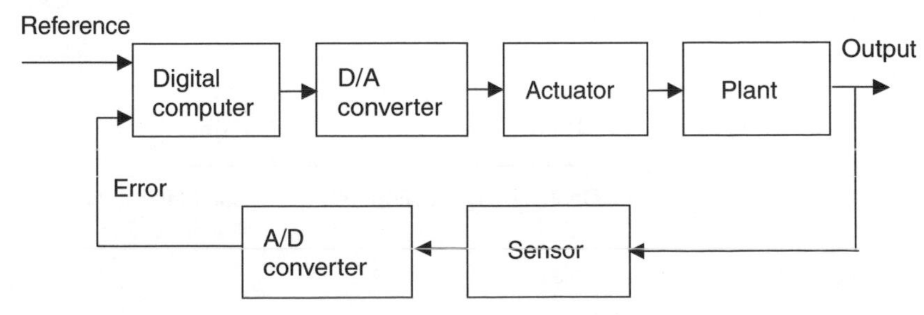


**Fig. 7.1** Digital control system

## 7.1 The sampling process

In continuous systems all system variables are continuous signals and their values are known or can be measured at all times. Digital control systems are sampled systems where the signals are known at some fixed periods, called the sampling period. For example, the computer reads the inputs at fixed time periods. The exact value of the inputs between these periods are not known. It also provides outputs at fixed time periods and again the exact values of the outputs between the sampling periods are not known.

The sampling process can be demonstrated by using an ideal sampler as shown in Fig. 7.2. A sampler is basically a switch that closes every T seconds for one instant of time. The input is r(t) and the output is denoted as $r^*(t)$. At sample time kT, the output can be written as

$$r^*(t) = r(kT)\delta(t - kT) \tag{7-1}$$

where $\delta$ is the impulse function defined as

$$\delta(t - kT) = \begin{cases} 1 & t = kT \\ 0 & t \neq kT \end{cases} \tag{7-2}$$

As shown in Fig. 7.3, if we sample a continuous signal r(t) at fixed intervals T, we obtain $r^*(t)$ which is the sum of impulses starting at $t = 0$ and having the amplitudes of t(kT). The output can be written as

$$r^*(t) = \sum_{k=0}^{\infty} r(kT)\delta(t - kT) \tag{7-3}$$

A D/A converter is used in Fig. 7.1 to convert a sampled signal to a continuous signal. The function of the D/A converter can be represented by a zero-order hold as shown in Fig. 7.4. The zero-order hold receives input r(kT) and holds the output constant for $kT = r < (k + 1)T$. The response of a zero-order hold to an impulse input is shown in Fig. 7.5.

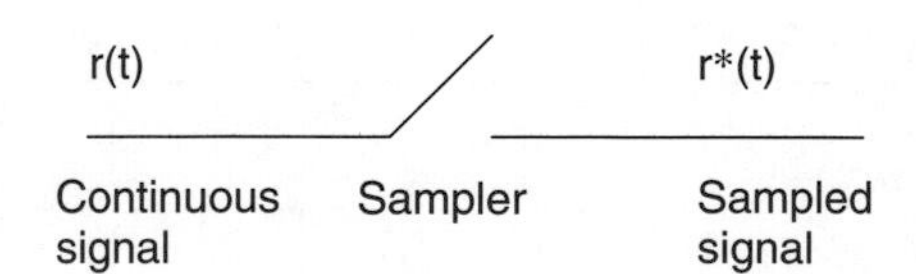


**Fig. 7.2** An ideal sampler

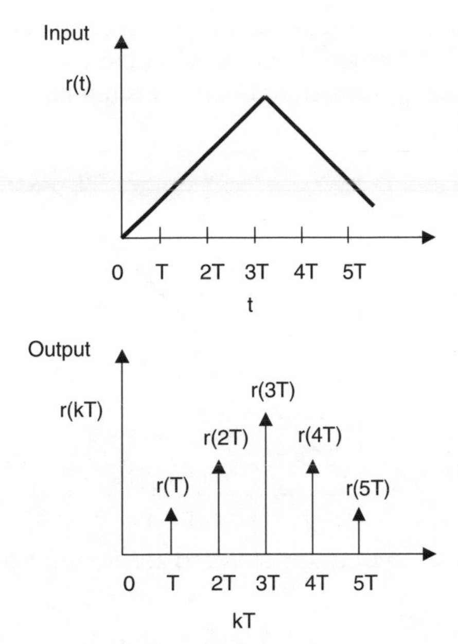


**Fig. 7.3** Input and output signals of the sampler

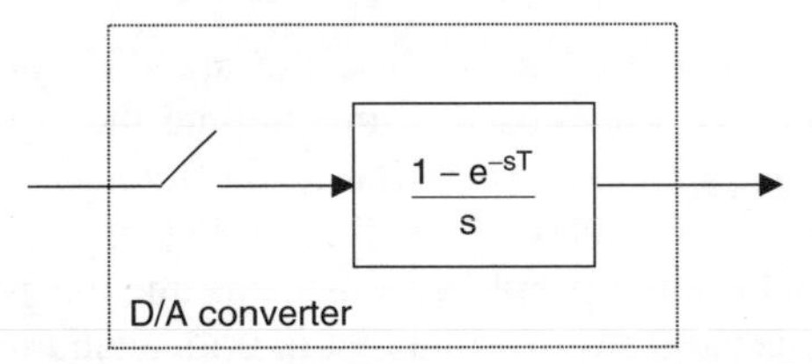


**Fig. 7.4** D/A converter as a sampler and a zero-order hold

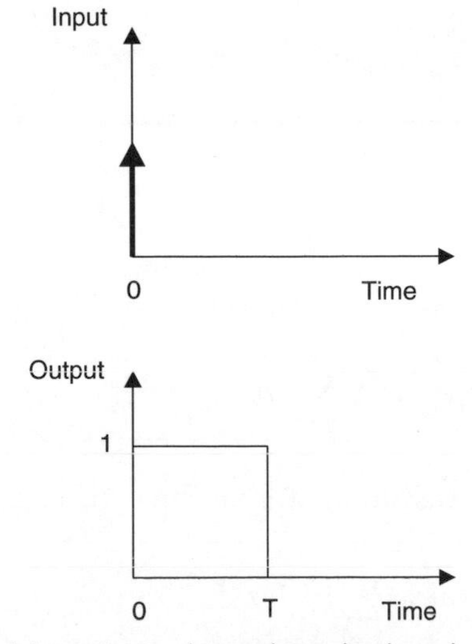


**Fig. 7.5** Zero-order hold response to an impulse input

An ideal sampler and a zero-order hold can accurately follow an input signal if the sampling period T is small compared to the changes in the signal. The response of a sampler and a zero-order hold to a ramp input is shown in Fig. 7.6.

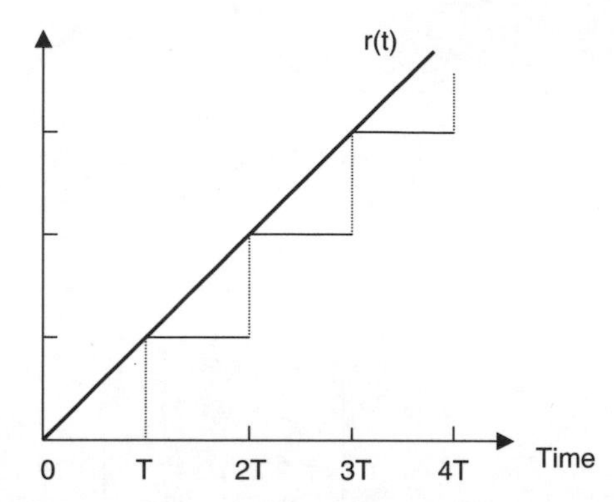


**Fig. 7.6** Response of a sampler and zero-order hold to ramp input

## 7.2 The z-transform

The z-transform is used to analyse the stability and the response of a digital control system. It is also used to manipulate the digital system block diagrams. z-transforms have properties similar to the Laplace transforms in the continuous domain and the symbol z being the analog of the symbol s. There are many similarities between the Laplace transforms and the z-transforms and a familiarity with the Laplace transform will assist in understanding and using the z-transform.

Using the Laplace transform we can re-write equation 7-3 for $t > 0$ as

$$L\{r^*(t)\} = \sum_{k=0}^{\infty} r(kT)e^{-ksT} \tag{7-4}$$

We can now define

$$z = e^{sT} \tag{7-5}$$

and then the z-transform is defined as

$$Z\{r(t)\} = Z\{r^*(t)\} = \sum_{k=0}^{\infty} r(kT)z^{-k} \tag{7-6}$$

In general, the z-transform of a function f(t) is defined as

$$Z\{f(t)\} = F(z) = \sum_{k=0}^{\infty} f(kT)z^{-k} \tag{7-7}$$

**Example 7.1**

Determine the z-transform of a unit step function.

**Solution 7.1**

A unit step function is defined as

$$f(t) = \begin{cases} 1 & t \geq 0 \\ 0 & t < 0 \end{cases}$$

Using equation 7-7

$$F(z) = \sum_{k=0}^{\infty} z^{-k} = 1 + z^{-1} + z^{-2} + z^{-3} + \cdots$$

using the identity

$$\frac{1}{1-x} = 1 + x + x^2 + x^3 + \cdots \qquad |x| < 1 \tag{7-8}$$

F(z) is now obtained

$$F(z) = \frac{1}{1 - z^{-1}} = \frac{z}{z - 1} \qquad |z^{-1}| < 1 \tag{7-9}$$

**Example 7.2**

Determine the z-transform of the exponential $f(t) = e^{-at}$ for $t \geq 0$

**Solution 7.2**

Using equation 7-7,

$$F(z) = \sum_{k=0}^{\infty} e^{-akT} z^{-k}$$

$$= \sum_{k=0}^{\infty} (ze^{aT})^{-k} = 1 + (ze^{aT}) + (ze^{aT})^{-2} + (ze^{aT})^{-3} + \cdots$$

using identity 7-8, we have

$$F(z) = \frac{1}{1 - (ze^{aT})^{-1}} = \frac{z}{z - e^{-aT}}$$

A table of some commonly used z-transforms is given in Table 7.1.

**Table 7.1** z-transforms of commonly used functions

| Time function | Laplace transform | z-transform |
|---|---|---|
| $1$ | $\frac{1}{s}$ | $\frac{z}{z-1}$ |
| $t$ | $\frac{1}{s^2}$ | $\frac{zT}{(z-1)^2}$ |
| $\frac{t^2}{2}$ | $\frac{1}{s^3}$ | $\frac{T^2z(z+1)}{(z-1)^3}$ |
| $\frac{t^3}{6}$ | $\frac{1}{s^4}$ | $\frac{T^3z(z^2+4z+1)}{(z-1)^4}$ |
| $e^{-at}$ | $\frac{1}{s+a}$ | $\frac{z}{z-e^{-aT}}$ |
| $te^{-at}$ | $\frac{1}{(s+a)^2}$ | $\frac{zTe^{-aT}}{(z-e^{-aT})^2}$ |
| $\sin at$ | $\frac{a}{s^2+a^2}$ | $\frac{z\sin aT}{z^2-2z\cos aT+1}$ |
| $\cos at$ | $\frac{s}{s^2+a^2}$ | $\frac{z(z-\cos aT)}{z^2-2z\cos aT+1}$ |
| $e^{-at}\sin bt$ | $\frac{b}{(s+a)^2+b^2}$ | $\frac{ze^{-aT}\sin bT}{z^2-2ze^{-aT}\cos bT+e^{-2aT}}$ |
| $e^{-at}\cos bt$ | $\frac{s+a}{(s+a)^2+b^2}$ | $\frac{z^2-ze^{-aT}\cos bT}{z^2-2ze^{-aT}\cos bT+e^{-2aT}}$ |
| $1-e^{-at}$ | $\frac{a}{s(s+a)}$ | $\frac{z(1-e^{-aT})}{(z-1)(z-e^{-aT})}$ |

## 7.2.1 Properties of the z-transform

The properties of the z-transform are given in this section without proof.

### Addition and subtraction

The z-transform of a sum of functions is equal to the z-transform of the overall function. That is,

$$Z[f_1(t) \pm f_2(t)] = F_1(z) \pm F_2(z)$$

**Multiplication by a constant**

The z-transform of a function multiplied by a constant is equal to the constant multiplied by the z-transform of the function:

$$Z[af(t)] = aF(z)$$

**Shift property I**

$$Z\{f(t+nT)\} = z^{n}\left[F(z) - \sum_{k=0}^{n-1} f(kT)z^{-k}\right]$$

**Shift property II**

$$Z\{e^{-at}f(t)\} = F(ze^{aT})$$

**Shift property III**

$$Z\{f(t-nT)\} = z^{-n}F(z)$$

**Initial value theorem**

Given that z-transform of f(t) is F(z), then

$$\lim_{k\to 0} f(kT) = \lim_{z\to\infty} F(z)$$

**Final value theorem**

$$\lim_{k\to\infty} f(kT) = \lim_{z\to 1}(z-1)F(z)$$

**Example 7.3**

The z-transform of a digital control system is

$$F(z) = \frac{z}{(z-0.6)(z-1)}$$

Calculate the final value of the system output.

**Solution 7.3**

Using the final value theorem

$$\lim_{k\to\infty} f(kT) = \lim_{z\to 1}(z-1)F(z)$$

$$= \left.\frac{(z-1)z}{(z-0.6)(z-1)}\right|_{z\to 1}$$

$$= \frac{1}{0.4} = 0.25$$

### 7.2.2 z-transform from the Laplace transform

It is important to note that although the z-transform of a system G(s) is denoted as G(z), it is not possible to obtain the system z-transform by simply substituting z for s in G(s). There are basically two methods to find the z-transform, given the Laplace transform. The first approach is to find the system time response by taking the inverse Laplace transform and then find the z-transform from the first principles, i.e. given a Laplace transform function G(s), find the time function g(t) using the inverse Laplace transform, and then find the equivalent z-transform G(z) using the z-transform properties. Some examples are given below to illustrate the method.

**Example 7.4**

Given

$$G(s) = \frac{10}{s+4}$$

Determine G(z).

**Solution 7.4**

We can first find the time function g(t) and then find the z-transform from Table 7.1.

The time function g(t) is found by taking the inverse Laplace transform. Thus,

$$g(t) = L^{-1}\{G(s)\} = 10\,e^{-4t}$$

from the z-transform table, Table 7.1,

$$G(z) = 10\frac{z}{z - e^{-4T}}$$

**Example 7.5**

Given

$$G(s) = \frac{1}{(s+a)(s+b)}$$

Determine G(z).

**Solution 7.5**

The time function g(t) is found using the inverse Laplace transform. Thus,

$$g(t) = L^{-1}\{G(s)\} = \frac{1}{b-a}(e^{-at} - e^{-bt})$$

from first principles (or using the z-transform tables),

$$G(z) = \sum_{k=0}^{\infty} \frac{1}{b-a}(e^{-akT} - e^{-bkT})z^{-k}$$

$$= \frac{1}{b-a}\Big[(1 + e^{-aT}z^{-1} + e^{-2aT}z^{-2} + \cdots)$$

$$- (1 + e^{-bT}z^{-1} + e^{-2bT}z^{-2} + \cdots)\Big]$$

$$= \frac{1}{b-a}\left[\frac{z(e^{-aT} - e^{-bT})}{(z - e^{-aT})(z - e^{-bT})}\right]$$

The second method to convert G(s) into G(z) is to assume that G(s) can be expressed as $G(s) = N(s)/D(s)$ and then use the following equation (the derivation of this equation is omitted):

$$G(z) = \sum_{n=1}^{p} \frac{N(x_n)}{D'(x_n)} \frac{1}{1 - e^{x_n T}z^{-1}} \tag{7-10}$$

where $x_n$, $n = 1, \ldots, p$ are the roots of the equation $D(s) = 0$ and $D' = \partial D/\partial s$. An example is given to illustrate the method.

**Example 7.6**

Given

$$G(s) = \frac{10}{s+4}$$

Determine the z-transform using the second method.

**Solution 7.6**

Writing G(s) as a numerator and denominator pair:

$$G(s) = \frac{N(s)}{D(s)}$$

where $N(s) = 10$, $D(s) = (s + 4)$, and $D'(s) = 1$. Also, $x_1 = -4$ and $p = 1$.

From equation 7-10,

$$G(z) = \sum_{n=1}^{p} \frac{N(x_n)}{D'(x_n)} \frac{1}{1 - e^{x_n T}z^{-1}} = \frac{10}{1} \frac{1}{1 - e^{-4T}z^{-1}}$$

$$= \frac{10z}{z - e^{-4T}}$$

**Example 7.7**

Given

$$G(s) = \frac{1}{s^2 + 3s + 2}$$

determine G(z) using the second method.

**Solution 7.7**

Again, writing G(s) as a numerator and denominator pair:

$$G(s) = \frac{N(s)}{D(s)}$$

where $N(s) = 1$, $D(s) = s^2 + 3s + 2 = (s + 1)(s + 2)$, and $D'(s) = 2s + 3$. Also, $x_1 = -1$, $x_2 = -2$, and $p = 2$.

From equation 7-10,

$$G(z) = \sum_{n=1}^{p} \frac{N(x_n)}{D'(x_n)} \frac{1}{1 - e^{x_n T} z^{-1}}$$

or,

$$G(z) = \frac{1}{-2+3} \frac{1}{1 - e^{-T} z^{-1}} + \frac{1}{-4+3} \frac{1}{1 - e^{-2T} z^{-1}}$$

$$= \frac{1}{1 - e^{-T} z^{-1}} - \frac{1}{1 - e^{-2T} z^{-1}}$$

$$= \frac{z(e^{-T} - e^{-2T})}{(z - e^{-T})(z - e^{-2T})}$$

# 7.3 Inverse z-transform

After the analysis and design of a digital control system is performed in the z domain, it is usually desired to determine the time response. The inverse z-transform converts from the z domain to the time domain. Since the z-transform defines a system only at the sampling instants, the inverse transformation gives the time values only at these instants, i.e. it is not possible to determine the time values in between the sampling instants. This is a limitation of the z-transform method. The inverse z-transform is denoted by $Z^{-1}$ and there are several methods that we can employ to determine the time function. Some popular methods are described here.

### 7.3.1 Power series method

This method involves dividing the denominator of F(z) into the numerator such that a power series of the form

$$F(z) = f_0 + f_1 z^{-1} + f_2 z^{-2} + \cdots$$

is obtained. The time function can then easily be determined. An example is given below to illustrate this method.

**Example 7.8**

Determine the time function f(t) for F(z) given by

$$F(z) = \frac{z}{z^2 - 3z + 3}$$

**Solution 7.8**

Using a long division, we obtain

$$\begin{array}{r|l} & z^{-1} + 3z^{-2} + 6z^{-3} + \cdots \\ z^2 - 3z + 3 & z \\ & \underline{z - 3 + 3z^{-1}} \\ & \quad 3 - 3z^{-1} \\ & \quad \underline{3 - 9z^{-1} + 9z^{-2}} \\ & \qquad 6z^{-1} - 9z^{-2} \\ & \qquad \underline{6z^{-1} - 18z^{-2} + 18z^{-3}} \\ & \qquad \ldots \end{array}$$

F(z) can be written as

$$F(z) = z^{-1} + 3z^{-2} + 6z^{-3} + \cdots$$

therefore,

$$f(0) = 0$$

$$f(1) = 1$$

$$f(2) = 3$$

$$f(3) = 6$$

or,

$$f(t) = \delta(t-1) + 3\delta(t-2) + 6\delta(t-3) + \cdots$$

### 7.3.2 Partial fraction expansion method

This method is similar to that employed with the inverse Laplace transform method where the function is expanded in partial fractions and tables of known z-transforms can be used to determine the inverse z-transform. Notice in the z-transform tables that a factor z appears in the numerator of the transforms. Thus, the partial fraction expansion should be performed on $F(z)/z$ so that the inverses can be read directly from the tables. Some examples are given below to illustrate the method.

**Example 7.9**

Function F(z) is given by

$$F(z) = \frac{z}{(z-1)(z-3)}$$

Determine the inverse z-transform using the partial expansion method.

**Solution 7.9**

We can re-write the function as

$$\frac{F(z)}{z} = \frac{1}{(z-1)(z-3)} = \frac{-1}{2(z-1)} + \frac{3}{2(z-3)}$$

Then

$$Z^{-1}\{F(z)\} = Z^{-1}\left[\frac{-z}{2(z-1)}\right] + Z^{-1}\left[\frac{3z}{2(z-3)}\right]$$

from the z-transform tables,

$$f(k) = -\frac{1}{2} + \frac{3}{2} \cdot 3^k$$

or,

$$f(t) = \delta(t) + \frac{5}{2}\delta(t-1) + \frac{11}{2}\delta(t-2) + \cdots$$

**Example 7.10**

Function F(z) is given by

$$F(z) = \frac{z}{(z-1)(z-2)(z-3)}$$

Determine the inverse z-transform using the partial expansion method.

**Solution 7.10**

We can re-write the function as

$$\frac{F(z)}{z} = \frac{1}{(z-1)(z-2)(z-3)} = \frac{1}{2(z-1)} - \frac{1}{(z-2)} + \frac{1}{2(z-3)}$$

Then

$$Z^{-1}\{F(z)\} = Z^{-1}\left[\frac{z}{2(z-1)}\right] - Z^{-1}\left[\frac{z}{(z-2)}\right] + Z^{-1}\left[\frac{1}{2(z-3)}\right]$$

From z-transform tables,

$$f(k) = f(k) = \frac{1}{2} - 2^k + \frac{3^k}{2}$$

## 7.4 The pulse transfer function

In this section we develop expressions for the z-transform of open-loop and closed-loop systems.

First of all we will look at some useful properties which will be used in the manipulation of block diagrams.

Consider the open-loop systems shown in Fig. 7.7. We will accept the following properties whose proofs are not given here:

1. Let $y(s) = G(s)u(s)$ then,

$$y(z) = y^*(s) = [G(s)u(s)]^* = G(z)u(z) \tag{7-11}$$

and

$$y^*(s) \neq G^*(s)u^*(s) \tag{7-12}$$

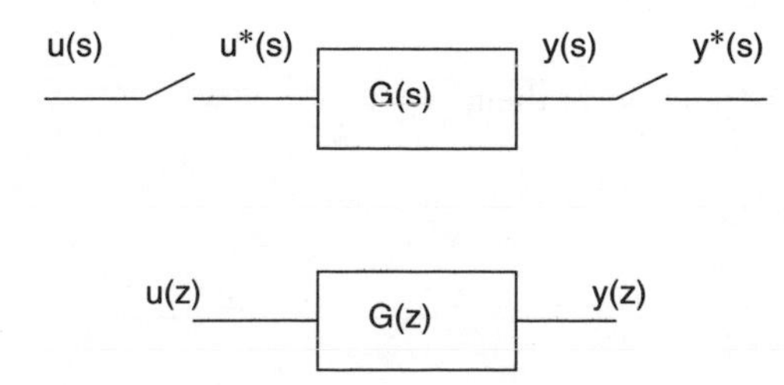


**Fig. 7.7** Open-loop system

This property tells us that the z-transform of a product is equal to the z-transform of the overall function formed by multiplying the individual functions.

2. Let $y(s) = G(s)u^*(s)$ then,

$$y^*(s) = [G(s)u^*(s)]^* = G^*(s)u^*(s) = G(z)u(z) \quad (7\text{-}13)$$

This property tells us that if at least one of the functions is sampled then the z-transform of the product is equal to the product of the z-transform of each function.

## 7.4.1 Open-loop digital control systems

Some example open-loop digital systems and their transfer functions are given in this section.

**Example 7.11**

Derive the transfer function of the open-loop system shown in Fig. 7.8.

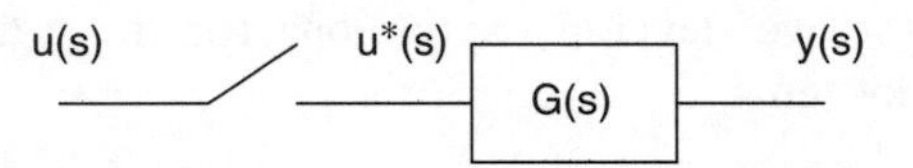


**Fig. 7.8** Open-loop system

**Solution 7.11**

With reference to Fig. 7.8,

$$y(s) = G(s)u^*(s)$$

using property in equation 7-13,

$$y^*(s) = [G(s)u^*(s)]^* = G^*(s)u^*(s)$$

or,

$$y(z) = G(z)u(z)$$

The transfer function is then,

$$\frac{y(z)}{u(z)} = G(z)$$

**Example 7.12**

Derive the transfer function of the open-loop system shown in Fig. 7.9.

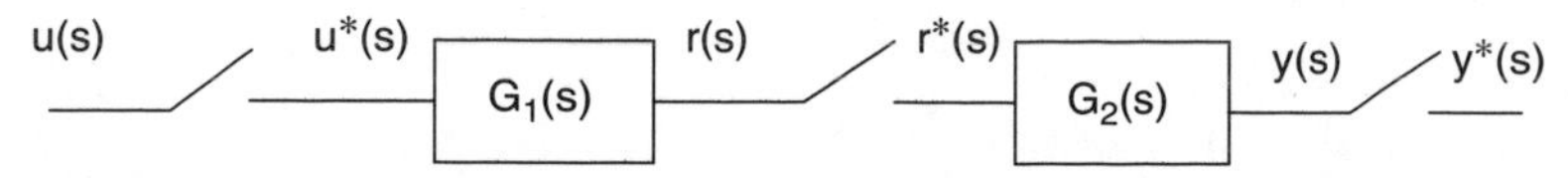


**Fig. 7.9** Open-loop system

**Solution 7.12**

With reference to Fig. 7.9,

$$y(s) = G_2(s)r^*(s)$$

or,

$$y^*(s) = G_2^*(s)r^*(s) \tag{7-14}$$

$$r(s) = G_1(s)u^*(s)$$

or,

$$r^*(s) = G_1^*(s)u^*(s) \tag{7-15}$$

Combining equations 7-14 and 7-15,

$$y^*(s) = G_2^*(s)G_1^*(s)u^*(s)$$

or,

$$y(z) = G_2(z)G_1(z)u(z)$$

and the transfer function is

$$\frac{y(z)}{u(z)} = G_2(z)G_1(z)$$

**Example 7.13**

Derive the transfer function of the open-loop system shown in Fig. 7.10.

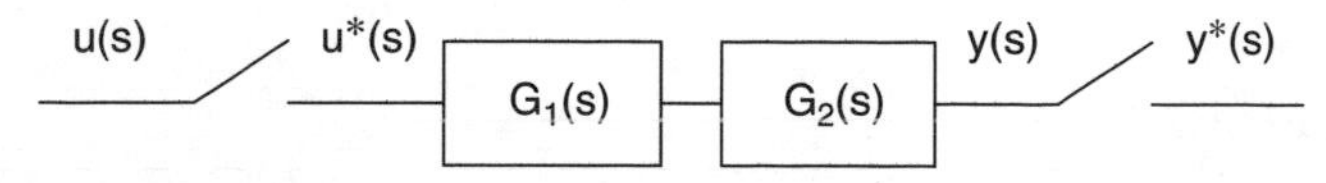


**Fig. 7.10** Open-loop system

**Solution 7.13**

With reference to Fig. 7.10,

$$y(s) = G_1(s)G_2(s)u^*(s)$$

or,

$$y^*(s) = [G_1(s)G_2(s)]^*u^*(s)$$

or,

$$y(z) = G_1G_2(z)u(z)$$

the transfer function can be written as

$$\frac{y(z)}{u(z)} = G_1G_2(z)$$

Notice that $G_1G_2(z)$ is determined by multiplying $G_1(s)$ and $G_2(s)$ and then finding the z-transform of the product.

## 7.4.2 Open-loop time response

In this section we give an example of how the time response of an open-loop digital control system can be calculated. An open-loop system block diagram is shown in Fig. 7.11, where $G_p(s)$ is the plant transfer function. G(s) is taken to be the product of the plant transfer function and a zero-order hold.

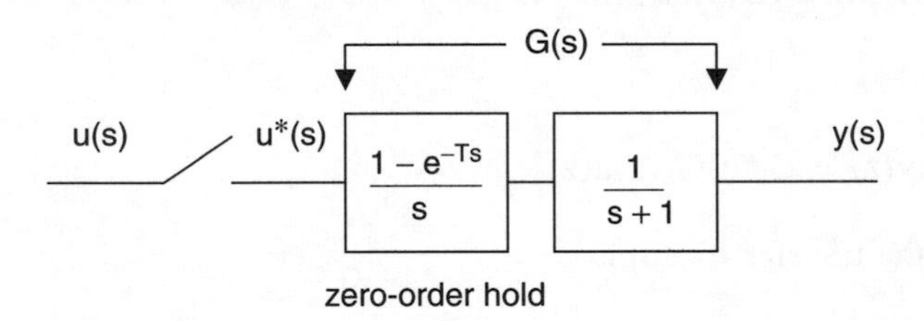


**Fig. 7.11** Example open-loop system

**Example 7.14**

Given the open-loop system shown in Fig. 7.11, with input r(t) a unit step function, determine the output of the system and draw a graph to show the output samples for the first several sampling intervals.

**Solution 7.14**

With reference to Fig. 7.11,

$$y(s) = G(s)u^*(s)$$

or

$$y^*(s) = G^*(s)u^*(s)$$

and

$$y(z) = G(z)u(z) \tag{7-16}$$

where G(z) is the plant transfer function, together with the zero-order hold function.

Now, we need to find the z-transform G(z) from a knowledge of G(s):

$$G(s) = \frac{1 - e^{-Ts}}{s(s+1)}$$

and

$$G(z) = (1 - z^{-1})Z\left[\frac{1}{s(s+1)}\right] \tag{7-17}$$

now, from z-transform tables,

$$Z\left[\frac{1}{s(s+1)}\right] = \frac{z(1 - e^{-T})}{(z-1)(z - e^{-T})}$$

Substituting in equation 7-17,

$$G(z) = (1 - z^{-1})\frac{z(1 - e^{-T})}{(z-1)(z - e^{-T})} = \frac{1 - e^{-T}}{z - e^{-T}} \tag{7-18}$$

In addition, the z-transform of unit step input u(t) is

$$u(z) = \frac{z}{z-1}$$

from equation 7-16, the output is

$$y(z) = G(z)u(z) = \frac{z(1 - e^{-T})}{(z-1)(z - e^{-T})}$$

$$= \frac{z}{z-1} - \frac{z}{z - e^{-T}}$$

From z-transform tables, the inverse transform yields

$$y(kT) = 1 - e^{-kT} \tag{7-19}$$

or, the output is

$$y(t) = (1 - e^{-T})\delta(t - T) + (1 - e^{-2T})\delta(t - 2T)$$

$$+ (1 - e^{-3T})\delta(t - 3T) + \cdots$$

Assuming T = 1 second, the first five output samples are:

$$y(0) = 0$$

$$y(1) = (1 - e^{-1})$$

$$y(2) = (1 - e^{-2})$$

$$y(3) = (1 - e^{-3})$$

$$y(4) = (1 - e^{-4})$$

$$y(5) = (1 - e^{-5})$$

...

Figure 7.12 shows the output of the system. Note that the output is only defined at the sampling instants.

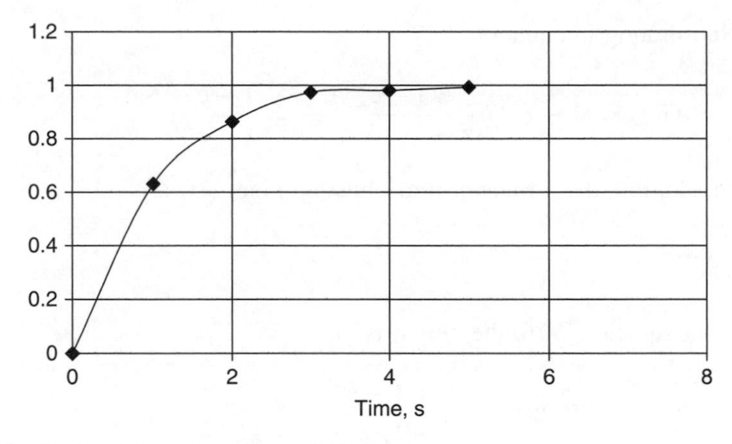


**Fig. 7.12** Output time response of the system

### 7.4.3 Closed-loop control systems

In this section we consider closed-loop digital control systems. Some example closed-loop systems and their transfer functions are given in this section.

**Example 7.15**

Derive the transfer function of the closed-loop system shown in Fig. 7.13.

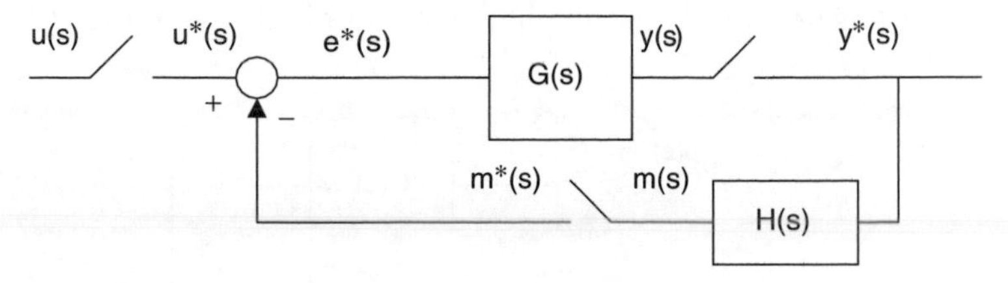


**Fig. 7.13** A closed-loop system

**Solution 7.15**

From the figure, we can write:

$$m(s) = H(s)y^*(s)$$

and,

$$m^*(s) = H^*(s)y^*(s)$$

$$e^*(s) = u^*(s) - m^*(s) = u^*(s) - H^*(s)y^*(s)$$

Therefore,

$$y(s) = G(s)e^*(s) = G(s)[u^*(s) - H^*(s)y^*(s)]$$

or,

$$y^*(s) = G^*(s)[u^*(s) - H^*(s)y^*(s)]$$

and

$$y^*(s)[1 + G^*(s)H^*(s)] = G^*(s)u^*(s)$$

The transfer function is then

$$\frac{y^*(s)}{u^*(s)} = \frac{G^*(s)}{1 + G^*(s)H^*(s)}$$

or in z-transform form

$$\frac{y(z)}{u(z)} = \frac{G(z)}{1 + G(z)H(z)}$$

The equivalent block diagram in z-transform is shown in Fig. 7.14.

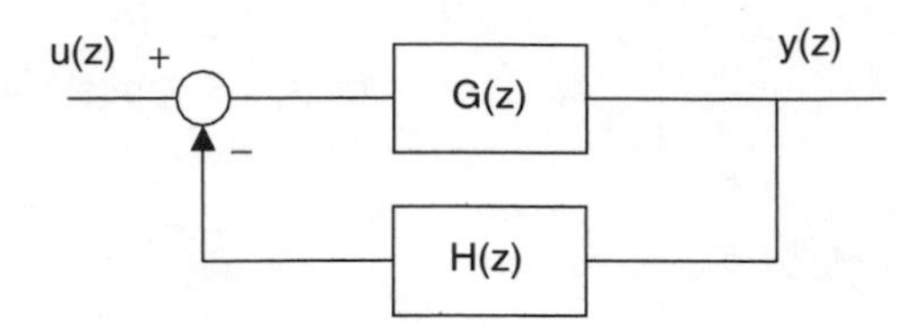


**Fig. 7.14** Block diagram in z-transform

### Example 7.16

Derive the transfer function of the closed-loop system shown in Fig. 7.15.

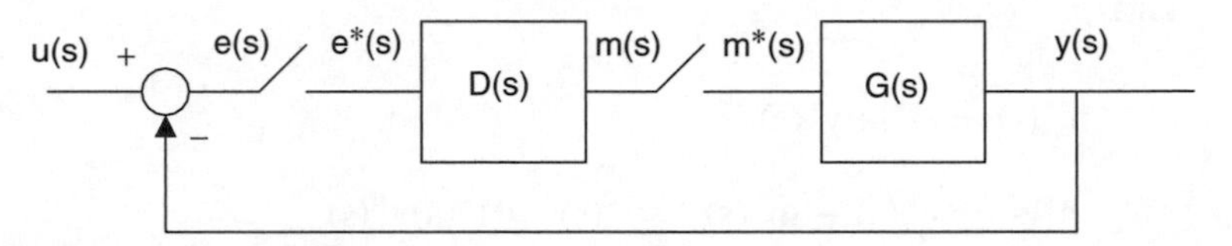


**Fig. 7.15** A closed-loop system

### Solution 7.16

From the figure, we can write:

$$y(s) = G(s)m^*(s)$$

or,

$$y^*(s) = G^*(s)m^*(s) \qquad (7\text{-}20)$$

Also,

$$m(s) = D(s)e^*(s)$$

or,

$$m^*(s) = D^*(s)e^*(s) \qquad (7\text{-}21)$$

From equations 7-20 and 7-21,

$$y^*(s) = G^*(s)D^*(s)e^*(s) \qquad (7\text{-}22)$$

Also,

$$e(s) = u(s) - y(s)$$

or,

$$e^*(s) = u^*(s) - y^*(s) \tag{7-23}$$

Substituting 7-23 into 7-22

$$y^*(s) = G^*(s)D^*(s)[u^*(s) - y^*(s)]$$

or,

$$y^*(s)[1 + G^*(s)D^*(s)] = G^*(s)D^*(s)u^*(s)$$

and

$$\frac{y^*(s)}{u^*(s)} = \frac{G^*(s)D^*(s)}{1 + G^*(s)D^*(s)}$$

Writing in z-transform format, the transfer function is

$$\frac{y(z)}{u(z)} = \frac{G(z)D(z)}{1 + G(z)D(z)}$$

Figure 7.16 shows the block diagram in z-transform format.

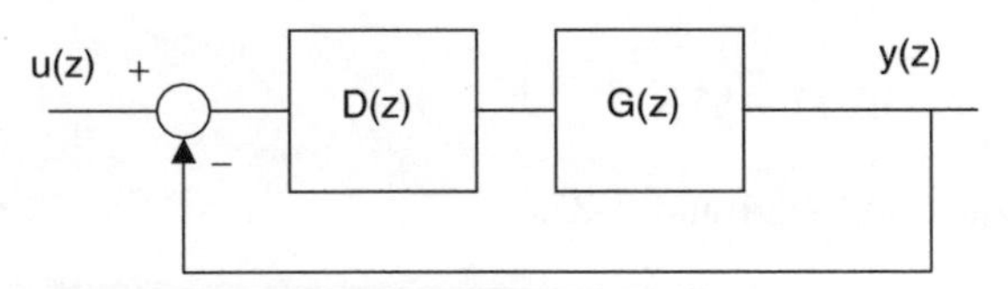


**Fig. 7.16** Block diagram in z-transform

## 7.4.4 Closed loop time response

In this section we give an example of how the time response of a closed-loop digital control system can be calculated. A closed-loop system block diagram is shown in Fig. 7.17, where $G_p(s)$ is the plant transfer function. G(s) is taken to be the product of the plant transfer function and a zero-order hold.

**Example 7.17**

Given the closed-loop system shown in Fig. 7.17, with input r(t) a unit step function, determine the output of the system and draw a graph to show the output samples for the first several sampling intervals. Assume that the sampling period is T = 1.

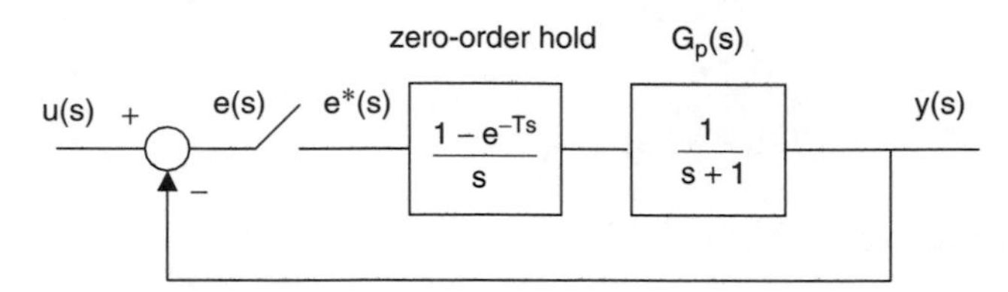


**Fig. 7.17** Example closed-loop system

**Solution 7.17**

With reference to Fig. 7.17,

$$G(s) = \frac{1 - e^{-sT}}{s(s+1)}$$

$$y(s) = G(s)e^*(s)$$

or,

$$y^*(s) = G^*(s)e^*(s) \tag{7-24}$$

Also,

$$e(s) = u(s) - y(s)$$

or,

$$e^*(s) = u^*(s) - y^*(s) \tag{7-25}$$

From equations 7-24 and 7-25,

$$y^*(s) = G^*(s)[u^*(s) - y^*(s)]$$

or,

$$y^*(s)[1 + G^*(s)] = G^*(s)u^*(s)$$

The transfer function is

$$\frac{y^*(s)}{u^*(s)} = \frac{G^*(s)}{1 + G^*(s)}$$

and in z-transform format

$$\frac{y(z)}{u(z)} = \frac{G(z)}{1 + G(z)} \tag{7-26}$$

Now, G(z) can be calculated from G(s):

$$G(z) = Z\left[\frac{1 - e^{-sT}}{s(s+1)}\right] = (1 - z^{-1})Z\left[\frac{1}{s(s+1)}\right] \tag{7-27}$$

From z-transform tables,

$$Z\left[\frac{1}{s(s+1)}\right] = \frac{z(1 - e^{-T})}{(z-1)(z - e^{-T})}$$

Substituting in equation 7-27,

$$G(z) = (1 - z^{-1})\frac{z(1 - e^{-T})}{(z-1)(z - e^{-T})} = \frac{1 - e^{-T}}{z - e^{-T}}$$

In addition, the z-transform of unit step input u(t) is

$$u(z) = \frac{z}{z-1}$$

From equation 7-26, the output is

$$y(z) = u(z)\frac{G(z)}{1 + G(z)}$$

Therefore,

$$y(z) = \frac{z}{(z-1)}\frac{(1 - e^{-T})}{(z - 2e^{-T} + 1)}$$

Since T = 1,

$$y(z) = \frac{z}{(z-1)}\frac{1 - e^{-1}}{(z - 2e^{-1} + 1)} = \frac{z}{(z-1)}\frac{1 - 0.3678}{(z - 2 \times 0.3678 + 1)}$$

$$= \frac{z}{(z-1)}\frac{0.6322}{(z - 0.2644)}$$

or,

$$\frac{y(z)}{z} = \frac{0.6322}{(z-1)(z - 0.2644)}$$

Using the partial fraction expansion technique,

$$\frac{y(z)}{z} = \frac{0.8594}{(z-1)} - \frac{0.8594}{(z - 0.2644)}$$

or,

$$y(z) = \frac{0.8594z}{(z-1)} - \frac{0.8594z}{(z-0.2644)}$$

The inverse z-transform is read from the tables as

$$y(k) = 0.8594 - 0.8594 \times 0.2644^k$$
$$= 0.8594(1 - 0.2644^k)$$

The output is

$$y(t) = 0.6321\delta(t-1) + 0.7993\delta(t-2) + 0.8435\delta(t-3) + 0.8552\delta(t-4)$$
$$+ 0.8582\delta(t-5) + 0.8591\delta(t-6) + \cdots$$

Notice that the final value of the output is 0.8594. The output response is plotted in Fig. 7.18.

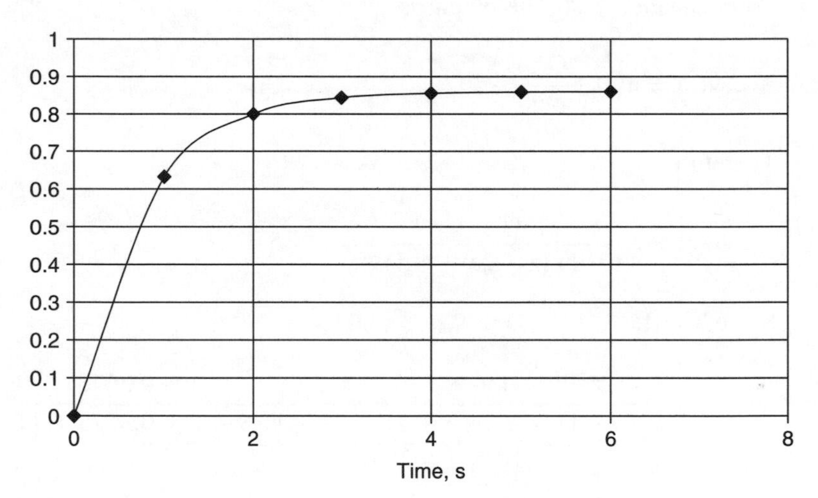


**Fig. 7.18** Output response

# 7.5 Exercises

1. Explain whether the following signals are discrete or continuous.

   a. Temperature in a room

   b. The display on a digital watch

c. The output of a speaker

d. An LCD display

2. Given the function

$$Y(s) = \frac{2}{(s+1)(s+2)}$$

use power series method and find Y(z) when T = 0.5 second.

3. Given the function

$$Y(s) = \frac{20}{s(s+1)(s+2)}$$

use the partial fraction expansion of Y(s) to find Y(z).

4. Find the values of y(kT) when

$$Y(z) = \frac{z}{z^2 - 3z + 3}$$

for $k = 0$ to 3.

5. Given the function

$$Y(s) = \frac{(s+3)}{s^2 + 3s + 2}$$

find Y(z) when T = 1 second.

6. Find the values of y(kT) when

$$Y(z) = \frac{z^3 + 2z^2 + 1}{z^3 - 2z^2 + z}$$

7. A digital control system has a plant transfer function

$$G_p(s) = \frac{120}{s^2 + 120}$$

Determine G(z) when a zero-order hold is used in the system. Assume that the sampling period is 0.1 second.

8. Find the z-transform of

$$Y(s) = \frac{(s+1)}{s^2 + 2s + 1}$$

when the sampling period is $T = 2$ seconds.

9. The open-loop system with a zero-order hold, as shown in Fig. 7.11, has a plant transfer function

$$G_p = \frac{2}{s(s+1)}$$

Calculate and plot the output response for the first four samples, when a unit step input is applied. Assume that the sampling period is $T = 1$ second.

10. A closed-loop system, as shown in Fig. 7.13, has the transfer functions

$$G(s) = \frac{2}{(s+1)}$$

$$H(s) = \frac{1}{s}$$

Calculate and plot the first four output samples when the sampling period is $T = 0.5$ second.

# Chapter 8

# Stability

The stability of a closed-loop transfer function

$$G(z) = \frac{N(z)}{D(z)}$$

depends on the location of the roots of equation $D(z) = 0$ which are the system closed-loop poles. For stability in the z-domain, all the roots of $D(z) = 0$ must lie inside a unit circle (compare this with the stability in the s-plane where all the roots of $D(s)$ must lie in the left-hand side of the s-plane).

Routh–Hurwitz stability criterion is probably the simplest method of determining the stability in the s-plane. Unfortunately, this method can not be used in the z-domain, but instead we can use similar methods, such as the Jury's test and the Shur–Cohn criterion. These stability tests are usually complicated, especially when high order systems are tested. We shall be looking at the Jury's test in the next section for systems of order 2 and 3.

## 8.1 Jury's stability test for small systems

Jury's stability test gets very complicated for high order systems. In this section, we describe the use of this test for small systems of order 2 or 3.

**Order 2**

Given the characteristic equation f(z) of a closed-loop system

$$f(z) = 1 + G(z)H(z) \tag{8-1}$$

where G(z) is the z-transform of the forward gain and H(z) is the z-transform of the feedback gain.

Let,

$$f(z) = a_0 + a_1 z + a_2 z^2 + \cdots$$

All roots of f(z) are inside the unit circle (i.e. the system is stable) if,

$$f(1) > 0, \qquad f(-1) > 0, \qquad |a_0| < a_2$$

**Example 8.1**

A unity gain feedback control system shown in Fig. 8.1 has the forward gain:

$$G(s) = \frac{1}{s(s+2)}$$

Use Jury's test to determine whether or not the closed-loop system is stable.

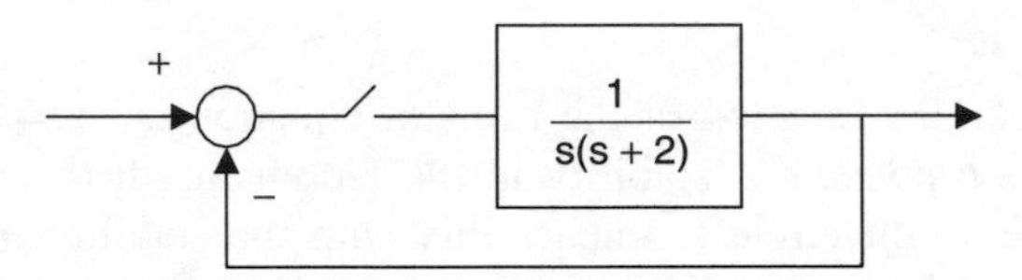


**Fig. 8.1** Closed-loop system for the example

**Solution 8.1**

The z-transform of G(s) is found as:

$$G(s) = \frac{1}{s(s+2)} = \frac{1}{2}\left(\frac{1}{s} - \frac{1}{s+2}\right)$$

From the z-transform tables,

$$G(z) = \frac{1}{2}\left[\frac{z}{z-1} - \frac{z}{z-e^{-2T}}\right]$$

or,

$$G(z) = \frac{1}{2}\frac{z(1-e^{-2T})}{(z-1)(z-e^{-2T})}$$

The characteristic equation is:

$$f(z) = 1 + G(z) = 1 + \frac{1}{2}\frac{z(1-e^{-2T})}{(z-1)(z-e^{-2T})}$$

or,

$$f(z) = 2z^2 - (1 + 3e^{-2T})z + 2e^{-2T}$$

We can now apply the three conditions of Jury's test:

i. $f(1) = 1 - e^{-2T} > 0$

this condition is satisfied

ii. $f(-1) = 3 + 5e^{-2T} > 0$

this condition is satisfied

iii. $a_0 = 2e^{-2T}$

$a_1 = -(1 + 3e^{-2T})$

$a_2 = 2$

and

$|a_0| < a_2$

this condition is satisfied for all positive values of T.

All the three conditions are satisfied and hence the system is stable with all the roots of the characteristic equation lying inside a unit circle.

**Order 3**

Given the characteristic equation

$$f(z) = a_0 + a_1 z + a_2 z^2 + \cdots$$

Jury's test states that all roots of f(z) will lie inside a unit circle if the following four conditions are satisfied:

$$f(1) > 0, \qquad f(-1) < 0, \qquad |a_0| < a_3$$

and

$$\left|\det\begin{bmatrix} a_0 & a_3 \\ a_3 & a_0 \end{bmatrix}\right| > \left|\det\begin{bmatrix} a_0 & a_1 \\ a_3 & a_2 \end{bmatrix}\right|$$

**Example 8.2**

The characteristic equation of a system is given by:

$$f(z) = 2z^3 + z^2 + 3z + 2$$

**Solution 8.2**

Applying the four conditions of Jury's test:

i. $f(1) = 8 > 0$

this condition is satisfied

ii. $f(-1) = -2 < 0$

this condition is satisfied

iii. $a_0 = 2$

$a_1 = 3$

$a_2 = 1$

$a_3 = 2$

and

$|a_0| < a_2$

this condition is not satisfied.

The system is not stable and the closed-loop system has poles on or outside the unit circle in the z-plane.

The stability of low order systems can easily be determined by examining the characteristic equation. An example is given below.

**Example 8.3**

The transfer function of a closed loop system is given by:

$$\frac{y(z)}{u(z)} = \frac{K(1 - e^{-T})}{z - e^{-T}(1 + K) + K}$$

where K is the forward gain. Determine the stability of this system.

**Solution 8.3**

The characteristic equation is:

$$z - e^{-T}(1 + K) + K = 0$$

or,

$$z = e^{-T} - K(1 - e^{-T})$$

The system is stable for small values of K and becomes unstable when

$$z = -1$$

or when

$$K = \frac{1 + e^{-T}}{1 - e^{-T}}$$

Assuming T = 1 second, the system will be unstable when

$$K = \frac{1 + e^{-1}}{1 - e^{-1}} = 2.17 \text{ or greater}$$

There are several other stability tests and design techniques such as the root-locus, frequency domain techniques, and so on. Root-locus is a graphical technique which shows the locations of the roots of the characteristic equation as the system gain is varied. The root-locus is a powerful tool for designing and analysing feedback control systems and its construction and use will be discussed in this section. Frequency domain techniques (e.g. Bode plots) are based on plotting the frequency response of the system using a logarithmic scale and then analysing the stability. Such techniques are unfortunately outside the scope of this book and the reader is referred to alternative books, specialized in the theory of digital control.

## 8.2 The root-locus technique

The stability and the transient behaviour of a closed-loop digital control system are directly related to the location of the closed-loop roots of the characteristic equation in the z-plane. It is usually possible to adjust one or more system parameters in order to locate the roots at desired points in the z-plane and thus obtain the required response. If, for example, the system gain varies, the pole positions of the closed-loop system will change as well. The set of these pole positions is called the root-locus of the system. The root-locus technique is widely used in both continuous, as well as in discrete time control systems.

Root–locus is a graphical technique for sketching the locus of the roots in the z-plane. It usually takes time to draw an accurate locus by hand and there are many computer programs (e.g. MATLAB) to draw the locus quickly and accurately. Some guidelines are given in this section to help the reader draw the root–locus of small systems without much effort.

Consider the general transfer function of a digital control system:

$$\frac{y(z)}{u(z)} = \frac{KG(z)}{1 + KG(z)H(z)}$$

The characteristic equation is:

$$1 + KG(z)H(z) = 0$$

The root–locus is drawn based on this characteristic equation. The following rules can be used to simplify the construction of a root–locus diagram:

- The root–locus starts at the poles of G(z)H(z) and terminates at the zeroes of G(z)H(z).
- The root–locus lies on a section of the real axis to the left of an odd number of poles and zeroes.
- The root–locus is symmetrical with respect to the horizontal axis.
- The root–locus may break away from the real axis and re-enter the real axis. The breakaway and the entry points are determined from the equation:

$$\frac{d[G(z)H(z)]}{dz} = 0$$

- The number of asymptotes is equal to the number of poles of G(z)H(z), $n_p$, minus the number of zeroes of G(z)H(z), $n_z$, with the asymptote angles given by:

$$\frac{(2k + 1)\pi}{(n_p - n_z)}, \quad k = 0, 1, 2, \ldots$$

- The asymptotes intersect the real axis at $\sigma$, where

$$\sigma = \frac{\sum \text{poles of } G(z)H(z) - \sum \text{zeroes of } G(z)H(z)}{n_p - n_z}$$

Some examples are given below to illustrate the construction of a root–locus diagram.

**Example**

The characteristic equation of a digital control system is given by:

$$1 + KG(z) = \frac{K(z+1)}{(z-1)^2}$$

construct the root–locus of this system and determine the values of K for which the system becomes unstable.

**Solution**

The poles of the characteristic equation are at $z = 1$ and the loci originate from these points. The zeroes are at $z = -1$ and at $z = -\infty$ and the loci terminate at these points. The region on the real axis between $-\infty$ and $-1$ are part of the loci since the number of poles and zeroes are odd in this section.

The breakaway points are obtained from:

$$\frac{d}{dz}[G(z)] = 0$$

which gives $z^2 + 2z - 3 = 0$ and the breakaway points are at $-3$ and $+1$.

$n_p = 2$ and $n_z = 1$ and therefore there is one asymptote with the angle $(2k+1)\pi/(n_p - n_z) = \pi$. The root–locus of the system is shown in Fig. 8.2. The locus starts at the double poles at $z = 1$ and then breakaway at $z = 1$. The locus re-enters the real axis at $z = -3$. One of the branches terminate at the zero at $z = -1$ and the other branch terminates at $-\infty$. It is clear from the locus that the system always has two roots outside the unit circle and is therefore always unstable for all positive values of K.

**Example**

The characteristic equation of a digital control system is given by:

$$1 + KG(z) = \frac{K(z+1)}{(z-1)(z-0.2)}$$

Draw the root–locus of the system and find the values of K for which the system is unstable.

**Solution**

The poles of the characteristic equation are at $z = 1$ and $z = 0.2$. The loci starts from these points and terminate at the zeroes which are at $z = -1$. The region of

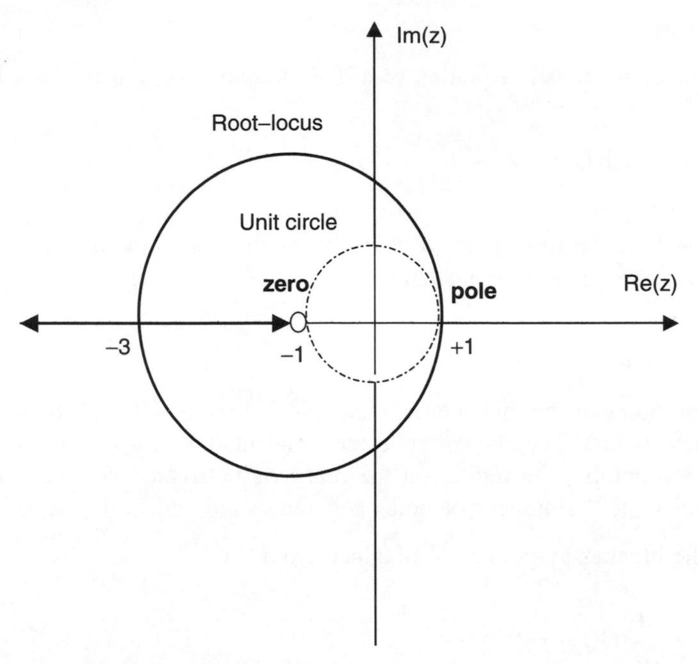


**Fig. 8.2** Root–locus for the example

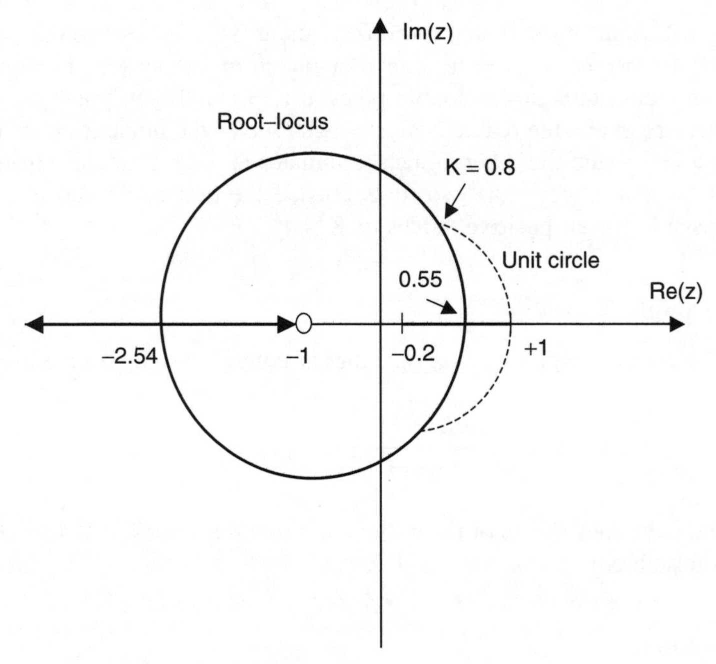


**Fig. 8.3** Root–locus for the example

real axis between $z = 1$ and $z = 0.2$ and also the region between $z = -\infty$ and $z = -1$ are part of the locus.

The breakaway points are obtained from:

$$\frac{d}{dz}[G(z)] = 0$$

which gives $z^2 + 2z - 1.4 = 0$ and the breakaway points are at $-2.54$ and $+0.55$.

$n_p = 2$ and $n_z = 1$ and therefore there is one asymptote with the angle $(2k + 1)\pi/(n_p - n_z) = \pi$. The root–locus of the system is shown in Fig. 8.3. The locus starts at the pole at $z = 1$ and $z = 0.2$. It then breaks away at $z = 0.55$. The locus re-enters the real axis at $z = -2.54$. One of the branches terminates at the zero at $z = -1$ and the other branch terminates at $-\infty$. The points of intersection of the locus with the unit circle may be found by graphical construction, or by the Jury stability test. The root–locus is on the unit circle at $K = 0.8$. Thus, the system is stable for $K < 0.8$.

## 8.3 Digital control algorithms

The aim of a digital control system is to achieve a particular control objective. This is done by programming a digital computer. Before a control algorithm is developed, it is essential to have an accurate mathematical model of the process to be controlled. The processes we have in the real-world are usually continuous time processes. In digital control, we create a mathematical model of these processes and then program a digital computer with the appropriate algorithm in order to move the process outputs to a desired state.

There are several methods that can be used when designing a digital control system:

- Model the process in continuous time to obtain the transfer function G(s). Then, find the corresponding z-transform G(z) and design a digital control algorithm.
- Model the process as a discrete time system and then design a digital control algorithm.
- Model the process in continuous time to obtain the transfer function G(s). Then, design a continuous controller using the well established s-plane techniques. Finally, discretize the resulting system into digital form.

This book is not a general digital control textbook, but is aimed for the professionals and students interested in temperature control, using digital computers.

As a result of this, only the techniques used mainly in temperature control will be discussed.

## 8.4 Temperature control using digital computers

Temperature control processes usually consist of some kind of device which provides heat output (e.g. a resistance heater) to the process. A temperature sensor is then used to sense the temperature and to provide feedback to the digital computer. Based on this feedback, the computer controls the heater and a fan in order to achieve the desired temperature in the plant. A simple digital temperature control system is shown in Fig. 8.4 where an analog sensor is used. Here, it is desired to keep the temperature of the liquid at a given set point. A resistance heater provides heat input to the liquid while the liquid is stirred continuously in order to achieve a uniform temperature in the liquid. A sensor detects the temperature and provides feedback to the digital computer. The digital computer runs a control algorithm and controls the heater in order to minimize or remove any error signals.

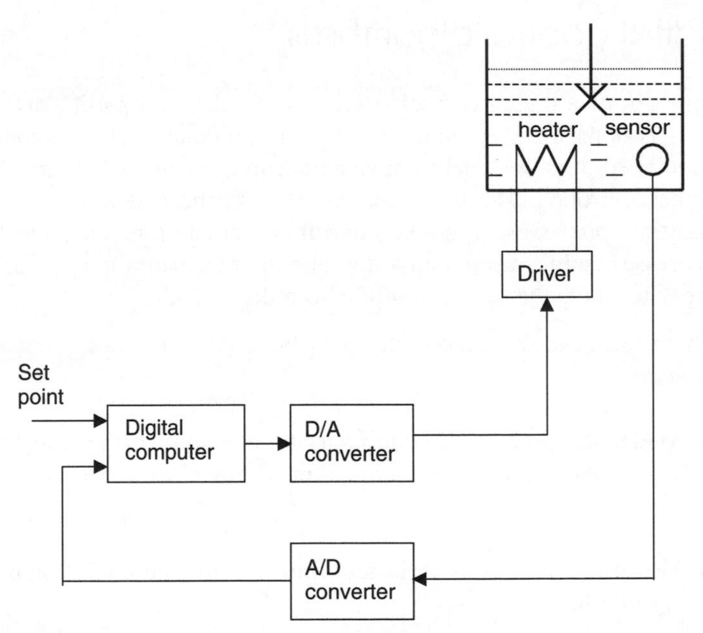


**Fig. 8.4** Digital computer temperature control

Most temperature control systems can be approximated to a single order system with a transportation delay (dead-time) and time constant which depends on the parameters such as the specific heat coefficient, temperature coefficient, volume, etc.

An example is given below which illustrates how the model of a simple temperature control system can be derived.

**Example 8.4**

Figure 8.5 shows an example temperature system where water at temperature $\theta_i$ enters a tank which is heated with an electric heater. The tank is insulated to reduce heat losses. Assume that the temperature of the water flowing out is $\theta$, and the ambient temperature is $\theta_a$. Also, assume that the water inside the tank is at a uniform temperature and there is no heat storage in the insulation. Derive the transfer function of this system.

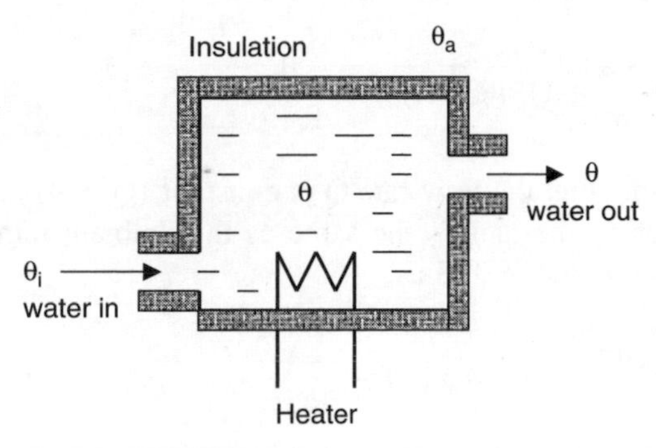


**Fig. 8.5** Example temperature system

**Solution 8.4**

At equilibrium, we can write the following equation:

$$q = q_{it} + q_{hl} + q_{is} - q_c \tag{8-2}$$

where

$q$ = rate of heat-flow of heating element
$q_{it}$ = rate of heat-flow into water in tank
$q_{hl}$ = rate of heat-flow by water leaving the tank
$q_{is}$ = rate of heat-flow through tank insulation
$q_c$ = rate of heat-flow into the tank by the entering cold water

Also,

$$q_{hi} = QP\theta$$

$$q_{it} = C\frac{d\theta}{dt}$$

$$q_c = QP\theta_i$$

and,

$$q_{is} = \frac{\theta - \theta_a}{R}$$

where

Q = water flow rate
P = specific heat coefficient of water
C = thermal capacitance of water
R = thermal resistance of the insulation

Combining the above equations, we can write

$$q = C\frac{d\theta}{dt} + QP(\theta - \theta_i) + \frac{\theta - \theta_a}{R} \tag{8-3}$$

If we assume that the flow rate Q is constant (Q = k) and the temperature of the water entering the tank is the same as the ambient temperature, i.e. $\theta_i = \theta_a$ we can re-write equation 8-3 as:

$$q = C\frac{d\theta}{dt} + (\theta - \theta_a)\left(kP + \frac{1}{R}\right) \tag{8-4}$$

If we let $\theta$ to be the temperature above the ambient temperature $\theta_a$, equation 8-4 becomes

$$q = C\frac{d\theta}{dt} + \left(kP + \frac{1}{R}\right)\theta \tag{8-5}$$

taking the Laplace transforms of both sides,

$$\frac{\theta(s)}{q(s)} = \frac{1}{Cs + kP + 1/R} \tag{8-6}$$

Letting

$$a = kP + \frac{1}{R}$$

$$b = \frac{1}{C}$$

the transfer function becomes

$$\frac{\theta(s)}{q(s)} = \frac{b}{s + ab} \tag{8-7}$$

Equation 8-7 describes a first-order system which gives the relationship between the water temperature and the rate of heat-flow of the heating element.

Some commonly used digital temperature control techniques are described in the following sections.

## 8.4.1 Bang-bang control of temperature

The simplest form of digital temperature control is to use a bang-bang (or on-off) type control. This form of control is used by almost all domestic thermostats. Here, as shown in Fig. 8.6, the temperature of the plant is compared to the desired temperature (set point) at each sampling time. If the plant temperature is higher than the desired temperature then the heater is turned off. Otherwise, the heater is turned on. In practice, the turn on temperature is made slightly less than the set point, and the turn off temperature is made slightly greater than the set point in order to prevent noise and wear-out from switching the heater rapidly on and off when the temperature is near the desired value. Figure 8.7 shows the block diagram of a bang-bang type control system.

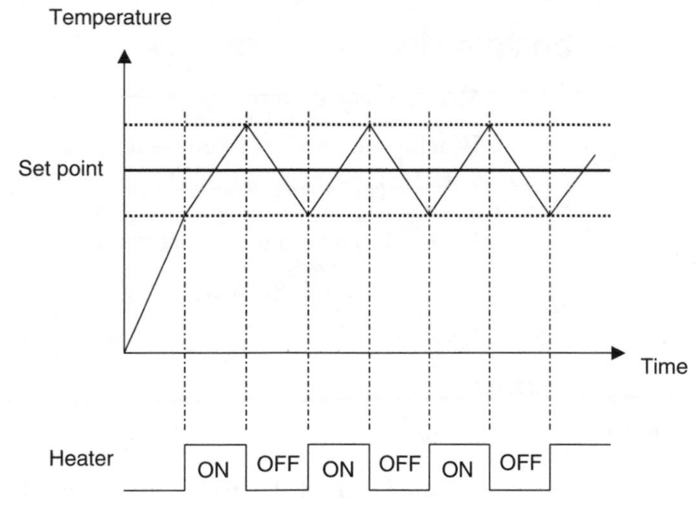


**Fig. 8.6** Bang-bang type temperature control

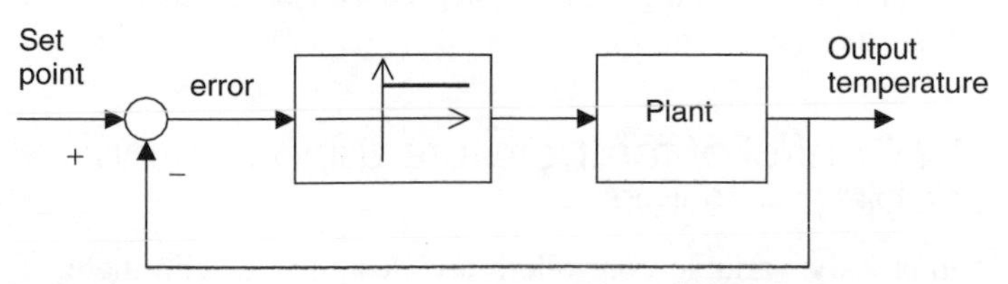


**Fig. 8.7** Block diagram of bang-bang control of temperature

In practice, a 2-way relay can be used as the controller and full heating power is applied when the measured temperature is low, and no heating power when the measured temperature is high.

The advantages of bang-bang type temperature control are:

- The controller is very simple and as such it is cheaper than other types of controllers.
- There is no need to have a detailed model of the plant.
- Setting the controller is very easy.
- This type of control is adequate in most industrial and commercial temperature control applications.
- The system can easily be made to be stable.

The bang-bang type temperature control can easily be implemented by a simple algorithm. The following PDL shows an implementation of this type of control.

```
BEGIN
    DO FOREVER
        Measure temperature
        IF temperature < Set point -d
            Turn ON heater
        ELSE IF temperature > Set point +d
            Turn OFF heater
        END IF
    ENDDO
END
```

There are special temperature control ICs which implement the bang-bang type control. For example, TC623 from Telcom Semiconductor Inc. contains a built-in temperature sensor and the device can be programmed by connecting two external resistors. The outputs of the device can be used to control a heater and a cooling fan.

### 8.4.2 Control of temperature using continuously variable controllers

Continuously variable controllers are designed to eliminate the cycling associated with bang-bang type of controllers. Such controllers measure the difference between the desired temperature and the actual temperature and use this value to determine how much power is to be fed to the heater. If the measured value is close to its desired value, less power is fed to the heater. As the measured value approaches its desired set point value, the power fed to the heater is progressively reduced.

One of the most commonly used controllers in temperature control is the Proportional-Integral-Derivative (PID or 3-term controller) controller. A PID controller looks at the current value of the error, the integral of the error over a time interval, and the current derivative of the error signal to determine not only how much a correction to apply, but the duration of the correction as well.

In continuous time domain, the PID algorithm has the general form:

$$u(t) = K_p e(t) + \frac{K_p}{T_i}\int_0^t e(t)\,dt + K_p T_d \frac{de(t)}{dt} \tag{8-8}$$

where e(t) is the error signal and u(t) is the control input to the process. $K_p$ is the proportional gain, $T_i$ is the integral time constant, and $T_d$ is the derivative time constant.

A proportional controller has the effect of reducing the rise time, but will never eliminate the steady-state error. Increasing the proportional gain will reduce the rise time, but it will also increase the overshoot. An integral controller has the effect of eliminating the steady-state error, but it may make the transient response worse. Too much integral action will cause large overshoots and an oscillatory behaviour. Large integral action also tend to decrease the rise time and increase the system settling time. A derivative controller has the effect of increasing the stability of the system, reducing the overshoot, and improving the transient response. Increasing the derivative action will decrease both the overshoot and the system settling time.

In s-domain, the PID controller can be written as

$$U(s) = K_p \left[1 + \frac{1}{T_i \cdot s} + TD \cdot s\right] E(s) \tag{8-9}$$

A PID controller has three parameters ($K_p, T_i, T_D$) which interact with each other and it can be a difficult task to tune these values in order to get the best performance.

The discrete form of the PID controller can be derived by finding the z-transform of equation 8-9:

$$U(z) = E(z) K_p \left[1 + \frac{T}{T_i(1 - z^{-1})} + T_D \frac{(1 - z^{-1})}{T}\right] \tag{8-10}$$

Equation 8-10 is also known as the *velocity* PID algorithm.

#### 8.4.2.1 Ziegler–Nichols PID tuning algorithm

Ziegler and Nichols suggested values for the PID parameters of a plant based on open-loop and closed-loop tests.

### OPEN-LOOP TUNING

According to Ziegler and Nichols, an open-loop process can be approximated by the transfer function

$$G(s) = \frac{Ke^{-sT_d}}{(1 + ST_1)} \tag{8-11}$$

where the coefficients K, $T_d$, and $T_1$ are found from a simple open-loop unit step response of the process as shown in Fig. 8.8.

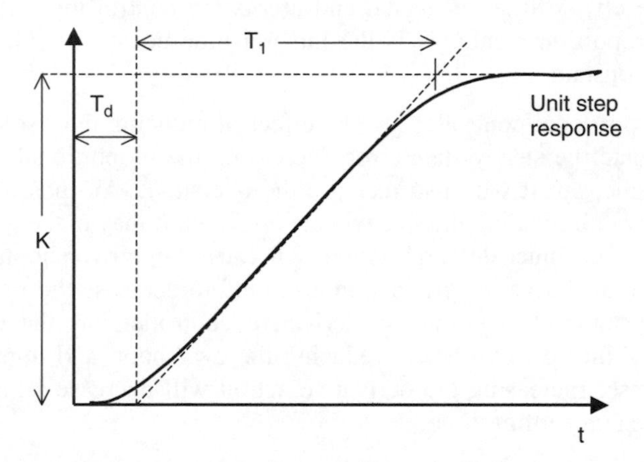


**Fig. 8.8** Finding parameters K, $T_d$, and $T_1$ of a process

Ziegler–Nichols then suggests the controller settings given in Table 8.1 in order to minimize the integral of absolute error after the application of a step change in the set point.

An example is given below to illustrate the method used.

**Example 8.5**

The open-loop unit step response of a thermal control system is shown in Fig. 8.9. Obtain the transfer function of this system and use the Ziegler–Nichols tuning algorithm to design a proportional and also a proportional-integral-derivative controller for the system. Draw the block diagrams of the system with both type of controllers.

**Table 8.1** Ziegler–Nichols settings

| Controller | $K_p$ | $T_i$ | $T_D$ |
|---|---|---|---|
| Proportional | $\frac{T_1}{KT_d}$ | | |
| Proportional + Integral | $\frac{0.9T_1}{KT_d}$ | $3.3T_d$ | |
| Proportional + Integral + Derivative | $\frac{1.2T_1}{KT_d}$ | $2T_d$ | $0.5T_d$ |

$K, T_d, T_1$ found from

$$G(s) = \frac{Ke^{-sT_d}}{(1 + sT_1)}$$

The controller is:

$$U(s) = K_p\left[1 + \frac{1}{T_i \cdot s} + T_D \cdot s\right]E(s)$$

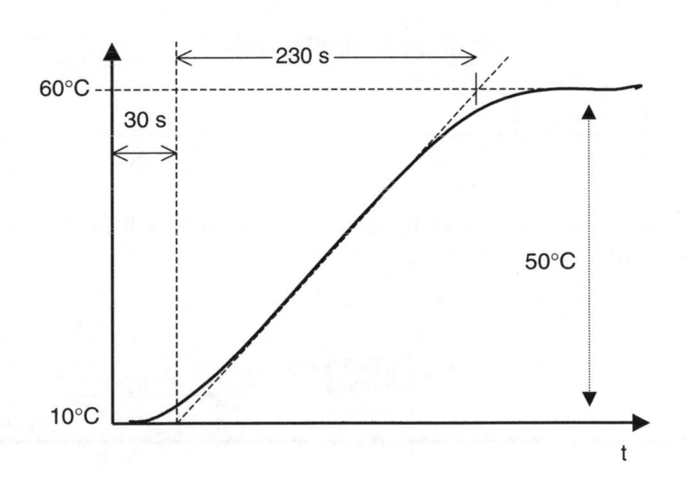


**Fig. 8.9** Unit step response of the thermal system

## Solution 8.5

From Fig. 8.9, the system parameters are obtained as:

$$K = 50^\circ C$$

$$T_d = 30\ \text{seconds}$$

$$T_1 = 230\ \text{seconds}$$

and the transfer function is

$$G(s) = \frac{50e^{-30\,s}}{(1 + 230\,s)}$$

**PROPORTIONAL CONTROL**

According to Table 8.1, the Ziegler–Nichols settings for a proportional controller are:

$$K_p = \frac{T_1}{KT_d}$$

Thus,

$$K_p = \frac{230}{50 \times 30} = 0.153$$

The transfer function of the controller is then

$$\frac{U(s)}{E(s)} = 0.153$$

and the block diagram of the closed-loop system with the controller is shown in Fig. 8.10.

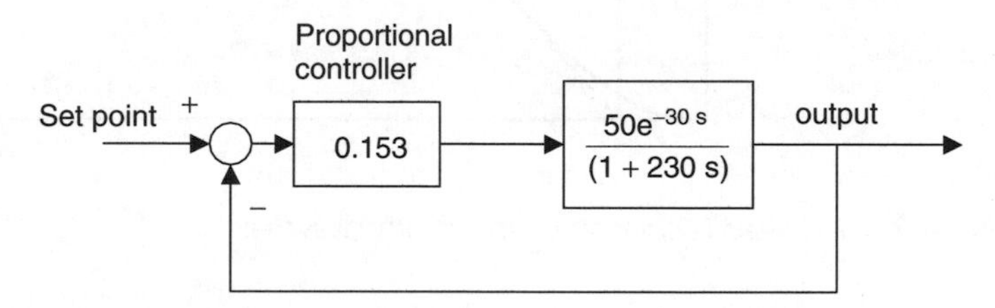


**Fig. 8.10** Block diagram of the system with proportional controller

**PID CONTROL**

According to Table 8.1, the Ziegler–Nichols settings for a proportional-integral-derivative controller are:

$$K_p = \frac{1.2T_1}{KT_d} \qquad Ti = 2T_d \qquad T_d = 0.5T_d$$

Thus,

$$K_p = \frac{1.2 \times 230}{50 \times 30} = 0.184 \qquad Ti = 2 \times 30 = 60 \qquad T_D = 0.5 \times 30 = 15$$

The transfer function of the required PID controller is:

$$\frac{U(s)}{E(s)} = 0.184\left[1 + \frac{1}{60\,s} + 15\,s\right]$$

or,

$$\frac{U(s)}{E(s)} = 0.184\left(\frac{900\,s^2 + 60\,s + 1}{60\,s}\right)$$

$$= \frac{165.6\,s^2 + 11.04\,s + 0.184}{60\,s}$$

and the block diagram of the closed-loop system with the controller is shown in Fig. 8.11.

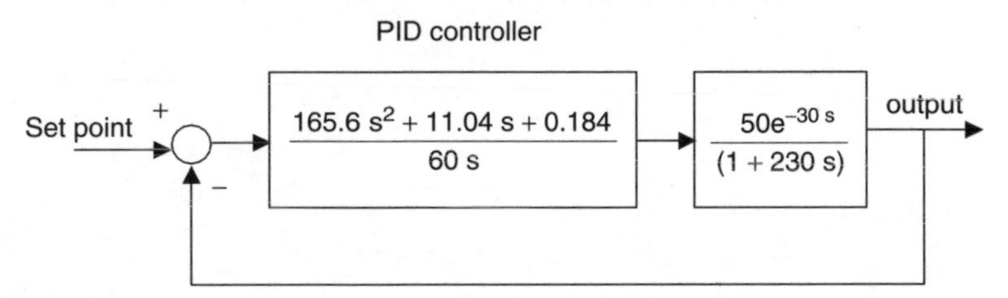


**Fig. 8.11** Block diagram of the system with PID controller

Note that the Ziegler–Nichols method does not give any guidance on what type of controller to use, but the best performance is usually obtained by using a PID type controller.

## CLOSED-LOOP TUNING

Ziegler–Nichols closed-loop tuning is based on closed-loop tests. The procedure is as follows (see Fig. 8.12):

- Disable any D and I action of the controller and leave only the P action
- Make a set point step test and observe the response
- Repeat the set point test with increased (or decreased) controller gain until a stable oscillation is achieved. This gain is called the "ultimate gain", $K_u$

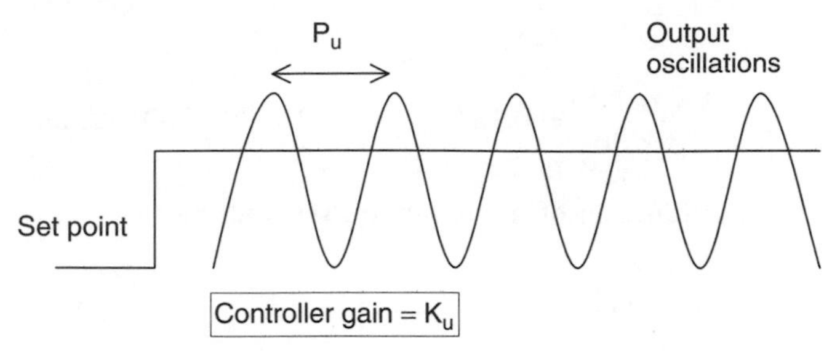


**Fig. 8.12** Ziegler–Nichols closed-loop tests

- Read the period of the steady oscillation. Let this be $P_u$
- Calculate the controller parameters according to the following formulas:

*Proportional and integral controller*:

$$K_p = 0.45K_u$$
$$T_i = P_u/1.2$$

*Proportional-integral-derivative controller*:

$$K_p = 0.6K_u$$
$$T_i = P_u/2$$
$$T_D = T_u/8$$

## 8.5 Digital realizations

A control algorithm represented by a z-transform transfer function must be realized using a computer. This is done using unit delays and constant multipliers. A given transfer function can be realized in several ways. Some of the most commonly used realization techniques are:

- Direct programming
- Serial programming
- Parallel programming
- Canonical programming

These realization techniques are described in the following sub-sections.

## 8.5.1 Direct programming

In this method, the transfer function is written in terms of the output function. An example is given below.

**Example 8.6**

Let the transfer function of a digital system be:

$$G(z) = \frac{Y(z)}{X(z)} = \frac{1 - 0.5z^{-1}}{(1 - z^{-1})(1 - 0.25z^{-1})}$$

Realize this transfer function using *direct programming*.

**Solution 8.6**

The transfer function can be written as:

$$Y(z) - 1.25z^{-1}Y(z) + 0.25z^{-2}Y(z) = X(z) - 0.5X(z)z^{-1}$$

after the inverse transformation,

$$y(kT) = 1.25y(kT - T) - 0.25y(kT - 2T) + x(kT) - 0.5x(kT - T)$$

we can now realize this equation as shown in Fig. 8.13.

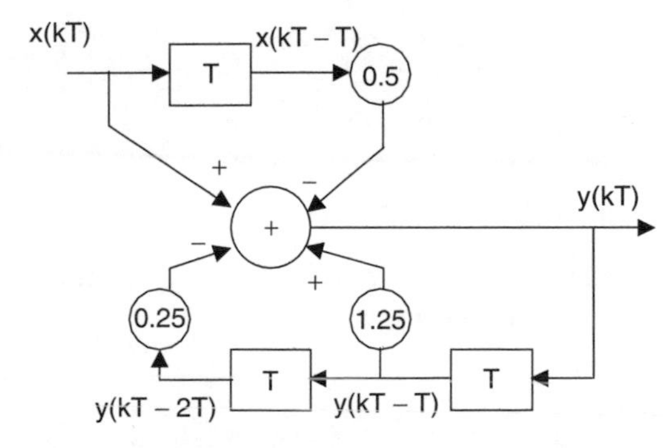


**Fig. 8.13** Direct realization of Example 8.6

## 8.5.2 Serial programming

In this method, the transfer function is considered as factors and each factor is realized separately and placed in cascade.

**Example 8.7**

Let the transfer function of a digital system be:

$$G(z) = \frac{Y(z)}{X(z)} = \frac{1 - 0.5z^{-1}}{(1 - z^{-1})(1 - 0.25z^{-1})}$$

Realize this transfer function using *serial programming.*

**Solution 8.7**

The transfer function can be written as:

$$G(z) = \frac{1 - 0.5z^{-1}}{(1 - z^{-1})(1 - 0.25z^{-1})} = (1 - 0.5z^{-1})\left[\frac{1}{(1 - z^{-1})}\frac{1}{(1 - 0.25z^{-1})}\right]$$

we can now realize this equation as shown in Fig. 8.14.

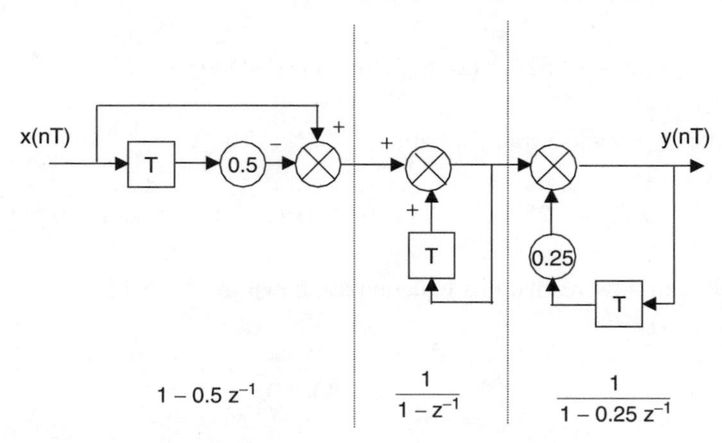


**Fig. 8.14** Serial realization of Example 8.7

## 8.5.3 Parallel programming

In parallel programming, the transfer function is split into a sum of modules and each module is connected in parallel.

**Example 8.8**

Let the transfer function of a digital system be:

$$G(z) = \frac{Y(z)}{X(z)} = \frac{1 - 0.5z^{-1}}{(1 - z^{-1})(1 - 0.25z^{-1})}$$

Realize this transfer function using *parallel programming.*

**Solution 8.8**

The transfer function can be written as:

$$G(z) = \frac{Y(z)}{X(z)} = \frac{0.66}{1 - z^{-1}} + \frac{0.33}{1 - 0.25z^{-1}}$$

we can realize this equation as in Fig. 8.15.

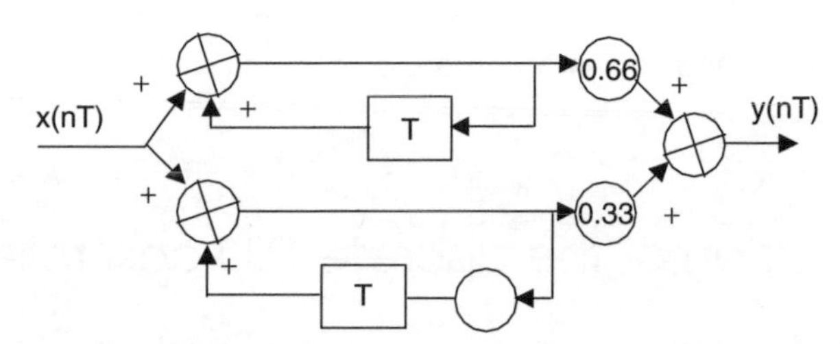


**Fig. 8.15** Parallel programming of Example 8.8

## 8.5.4 Canonical programming

In canonical programming we find the difference equations and then realize the transfer function based on these difference equations. Canonical programming is widely used in realizing digital controllers because of its simplicity and the ease of developing an algorithm.

**Example 8.9**

Let the transfer function of a digital system be:

$$G(z) = \frac{Y(z)}{X(z)} = \frac{1 - 0.5z^{-1}}{(1 - z^{-1})(1 - 0.25z^{-1})}$$

Realize this transfer function using *canonical programming*.

**Solution 8.9**

The transfer function can be written as:

$$y(kT) = x(kT) + [-0.5x(kT - T) + 1.25y(kT - T)] - 0.25y(kT - 2T)$$

we can realize this equation as in Fig. 8.16.

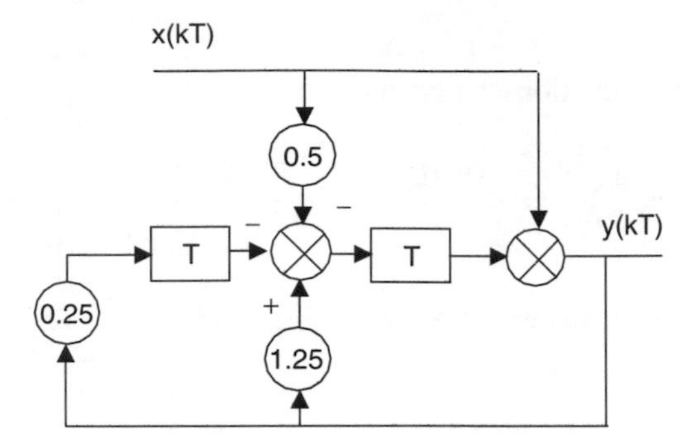


**Fig. 8.16** Canonical programming of Example 8.9

## 8.6 Realization of the discrete PID controller

The discrete PID controller was given in equation 8-9 as:

$$U(z) = E(z)K_p\left[1 + \frac{T}{T_i(1 - z^{-1})} + T_D\frac{(1 - z^{-1})}{T}\right] \tag{8-10}$$

This controller is usually realized using parallel programming since equation 8-10 is already in a form which can be realized easily using parallel programming.

Rewriting equation 8-10 as

$$\frac{U(z)}{E(z)} = a + \frac{b}{1 - z^{-1}} + c(1 - z^{-1}) \tag{8-12}$$

where

$$a = K_p \qquad b = \frac{K_pT}{T_i} \qquad c = \frac{K_pT_D}{T} \tag{8-13}$$

Equation 8-12 can be realized as shown in Fig. 8.17.

In reference to Fig. 8.17, the following equations can be written. Notice that p(kT) and q(kT) are temporary variables used in the equations:

$$p(kT) = be(kT) + p(kT - T)$$

$$q(kT) = ce(kT) - ce(kT - T)$$

$$u(kT) = p(kT) + aek(t) + q(kT)$$

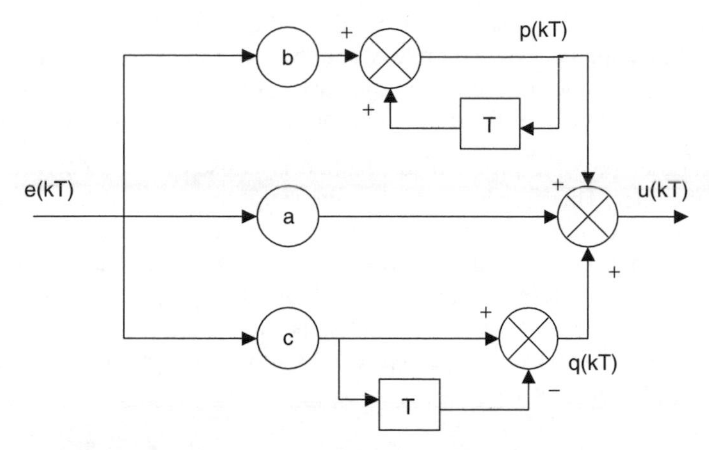


**Fig. 8.17** Parallel realization of the PID controller

We can now develop the following PDL to implement the PID algorithm (note that a, b, c are the PID parameters as defined in equation 8-13):

**BEGIN**

**DO FOREVER**

Get set point: r(kT)

Get system output: y(kT)

Calculate error: e(kT) = r(kT) − y(kT)

Calculate I term: p(kT) = be(kT) + p(kT−T)

Calculate D term: q(kT) = ce(kT) − ce(kT−T)

Calculate PID output: u(kT) = p(kT) + ae(kT) + q(kT)

Send control to actuator

Save variables: p(kT−T) = p(kT)

e(kT−T) = e(kT)

Wait for next sample

**ENDDO**

**END**

## 8.7 Problems with the standard PID controller

One of the practical problems when the standard form of the PID controller used is known as the "*Integral windup*" which can cause long periods of overshoot in the controlled response. Integral windup arises because of an overflow in the digital computer. For example, when the error signal is integrated for long periods of time, an overflow situation may arise. One way to avoid integral windup is to limit the signals in the digital computer to a maximum and minimum.

Another possible problem in practice is caused by the derivative action of the controller. This may happen when the set point changes sharply, causing the error signal to change. Under such a situation, the derivative term can give a "*kick*" in the output, known as the "*derivative kick*". This is avoided in practice by moving the derivative term to the feedback loop. The proportional term may also cause a sudden kick in the output and it is also common to move the proportional term to the feedback loop. Figure 8.18 shows the practical realization of the PID controller in practice.

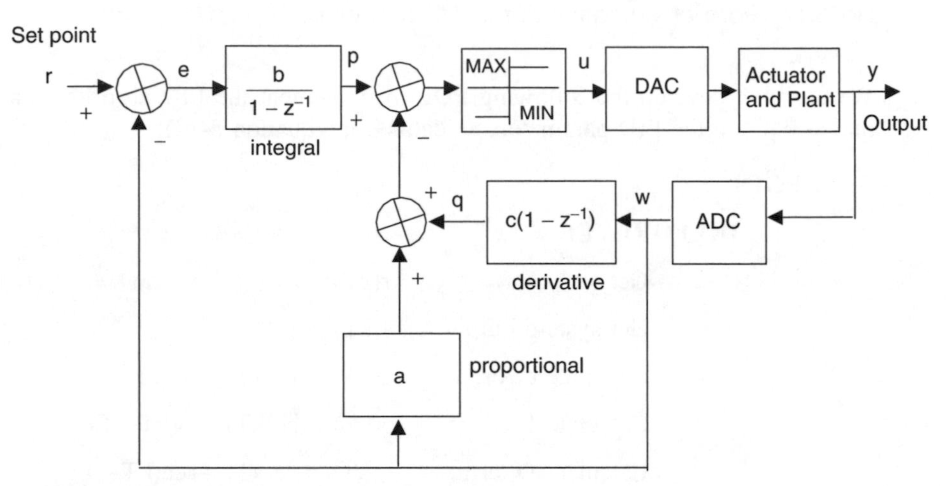


**Fig. 8.18** Practical realization of the PID controller

Based on Fig. 8.18, we can develop the following PDL for the practical implementation of a PID controller by a digital computer:

**BEGIN**

**DO FOREVER**

Get set point: r(kT)

Get system output: w(kT)

Calculate error: $e(kT) = r(kT) - w(kT)$

Calculate I term: $p(kT) = be(kT) + p(kT-T)$

Calculate D term: $q(kT) = cw(kT) - cw(kT-T)$

Calculate PID output: $u(kT) = p(kT) + aw(kT) + q(kT)$

```
            IF u(kT) > MAX
                    p(kT) = p(kT−T) + MAX − u(kT)
                    u(kT) = MAX
            ELSE IF u(kT) < MIN
                    p(kT) = p(kT−T) + MIN − u(kT)
                    u(kT) = MIN
            END IF
            Send control to actuator
            Save variables:      p(kT−T) = p(kT)
                                 w(kT−T) = w(kT)
            Wait for next sample
      ENDDO
END
```

## 8.8 Choosing a sampling interval

Whenever a digital control system is designed, a suitable sampling interval must be chosen. The simple choice is to sample as fast as possible. However, an unnecessarily fast sampling is a waste of resources. For example, fast sampling requires more expensive A/D converters. On the other hand, if the sampling rate is too low, then signal loss will occur.

There are many empirical rules for the selection of the sampling rate. The following is a guideline for the choice of a minimum sampling rate:

- If the closed-loop system is required to have a natural frequency of $w_n$, then choose the system sampling frequency as:

  $$w_s > 10w_n$$

  where $w_s$ is the sampling frequency, i.e. $w_s = \frac{2\pi}{T}$

  This is the same as choosing $T < T_s/10$ where $T_s$ is the closed-loop system settling time.

- If the system has a dominant time constant $T_{dtc}$ then choose the sampling interval T as:

  $$T < \frac{T_{dtc}}{10}$$

- If the system has the Ziegler–Nichols open-loop response as given by equation 8.12, then choose the sampling time T as:

$$T < \frac{T_1}{4}$$

## 8.9 Exercises

1. Investigate the stability of following transfer functions by Jury's test:

   a. $G(z) = \dfrac{1}{z^3 + 3z^2 + 2z + 0.5}$

   b. $G(z) = \dfrac{1}{z^3 - 4z^2 + 3z - 2}$

2. A closed-loop system has the characteristic equation:

$$\frac{y(z)}{u(z)} = \frac{3K(1 - e^{-T})}{z - e^{-T}(2 + K) + K}$$

   determine the stability of this system.

3. Explain the difference between a bang-bang type controller and a continuous controller.

4. Derive an equation for the discrete form of the PID controller. Draw the block diagram of this controller.

5. The open-loop unit step response of a thermal system is as shown in Fig. 8.19. Find the PID parameters of this system using the Ziegler–Nichols tuning method. Draw the block diagram of the final system.

6. Explain the Ziegler–Nichols closed-loop tuning method.

7. Realize the following transfer functions using direct programming:

   a. $G(z) = \dfrac{Y(z)}{X(z)} = \dfrac{1 - 0.5z^{-1}}{(1 - 0.5z^{-1})(1 - 0.8z^{-1})}$

   b. $G(z) = \dfrac{Y(z)}{X(z)} = \dfrac{2(1 - z^{-1})}{(1 - 0.2z^{-1})(1 - 0.4z^{-1})}$

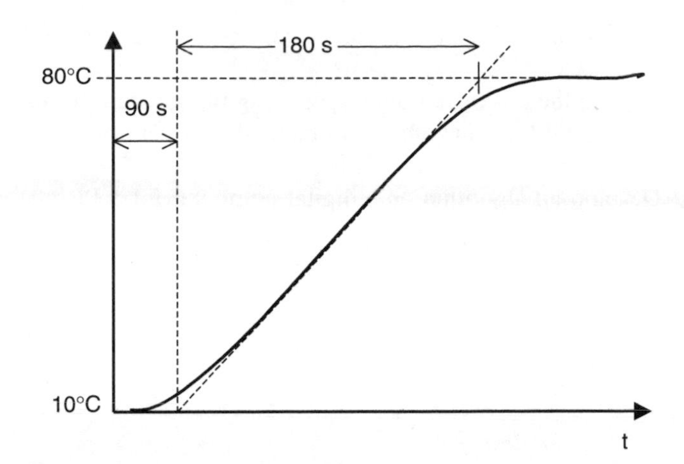


**Fig. 8.19** System response for Exercise 5

8. Realize the following transfer functions using parallel programming:

a. $G(z) = \dfrac{Y(z)}{X(z)} = \dfrac{z^{-1}}{(1 - 0.5z^{-1})(1 - 0.25z^{-1})}$

b. $G(z) = \dfrac{Y(z)}{X(z)} = \dfrac{2z^{-1}}{(1 - 2z^{-1})(1 - 4z^{-1})}$

c. $G(z) = \dfrac{Y(z)}{X(z)} = \dfrac{z}{z^3 + 2z^2 - 4z + 0.5}$

9. Realize the following transfer functions using serial programming:

a. $G(z) = \dfrac{Y(z)}{X(z)} = \dfrac{z^{-1}}{(1 - 2z^{-1})(1 - 4z^{-1})(1 - 3z^{-1})}$

b. $G(z) = \dfrac{Y(z)}{X(z)} = \dfrac{z^{-1}(1 - z^{-1})}{(1 - 0.5z^{-1})(1 - 0.25z^{-1})(1 - 0.1z^{-1})}$

10. Realize the following transfer functions using canonical programming:

a. $G(z) = \dfrac{Y(z)}{X(z)} = \dfrac{z}{z^3 + 2z^2 - 4z + 2}$

b. $G(z) = \dfrac{Y(z)}{X(z)} = \dfrac{2z}{z^4 + 2z^3 - 4z^2 + 0.5z - 2.5}$

11. Realize the PID controller using canonical programming.

12. Explain the practical problems of using the standard forms of PID controllers. Draw the block diagram of a practical PID controller.

13. Develop an algorithm on a digital computer for a PI type controller.

# Chapter 9

# Case Study: Temperature Control Project

The aim of this chapter is to show how a complete temperature control system can be designed from the first principles. The design of a microcontroller based control system is described from modelling to the control of the process.

## 9.1 Overview

An electrical water heater is considered as the example process. The block diagram of the process is shown in Fig. 9.1. An electric heater is used to heat-up the temperature of water in a tank and the aim is to keep the temperature at the desired value. As shown in the figure, the water temperature is sensed using an analog

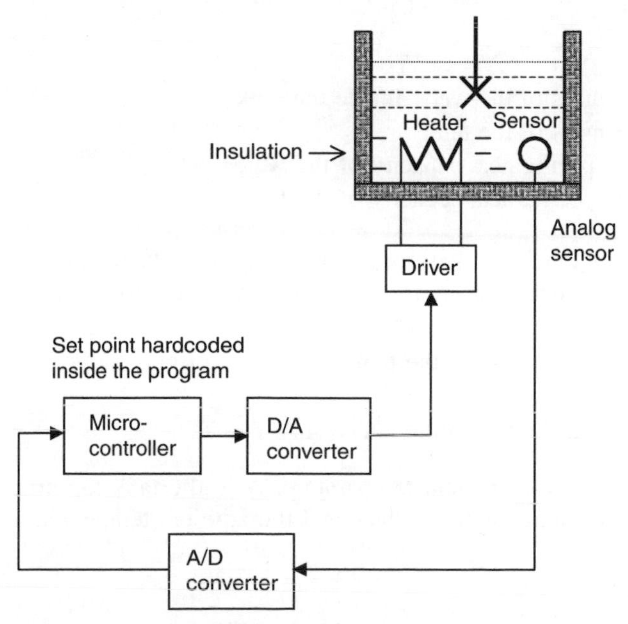


**Fig. 9.1** The temperature process

integrated circuit sensor (LM35DZ). The sensor output is converted into digital form and is compared with a stored desired temperature to form an error signal.

A PI and a PID type controller algorithms will be implemented by the microcontroller in order to achieve the desired output. The output of the microcontroller is converted into analog form and this signal is used as an input to a driver circuit which provides power to the heater element.

A PIC16F877 type microcontroller is used in this project. The reason for using this microcontroller is because of its A/D converter input and also Pulse Width Modulated (PWM) output, which is used to drive the heater element. PIC16F877 also has a large memory which is needed to implement floating-point arithmetic and the PI or the PID algorithm.

## 9.2 The mathematical model

The design of the controller will be based on the Ziegler–Nichols open-loop tuning method and thus an accurate mathematical model of the system is not normally required. It is however important to see what type of system we have as this may effect the decisions we make about the system.

### 9.2.1 Mathematical model of the tank

We can write the following heat-balance equation for the tank:

*heat input to the system = heat increase in the system + heat losses*

Let,

m1 = mass of the water inside the tank
m2 = mass of the tank
c1 = specific heat capacity of the water
c2 = specific heat capacity of the tank

Ignoring the heat loss through the walls of the tank and the heat capacities of the heater element and the mixer, we can write the following equations:

$$\text{Heat increase in the tank} = (m_1c_1 + m_2c_2)\frac{dT}{t}$$

$$\text{Heat loss from the tank} = hA(T - T_a)$$

Where, $T_a$ the ambient temperature, A is the tank top area and h is a constant which depends on the surface and the ambient temperature.

Thus,

$$E = (m_1c_1 + m_2c_2)\frac{dT}{t} + hA(T - T_a) \quad (9\text{-}1)$$

If we assume that the ambient temperature is constant, and let

$$T_q = T - T_a$$

we can write equation 9-1 as:

$$E = (m_1c_1 + m_2c_2)\frac{dT_q}{t} + hAT_q$$

or, letting $k_1 = m_1c_1 + m_2c_2$ and $k_2 = hA$, and taking the Laplace transforms,

$$\frac{T_q(s)}{E(s)} = \frac{1}{sk_1 + k_2} \tag{9-2}$$

Equation 9-2 describes a first-order system with time constant $k_1/k_2$. Temperature control systems always exhibit a transportation delay since it takes a finite time for the temperature of the medium to rise. The transportation delay time and the system parameters $k_1$ and $k_2$ will be determined from the step response tests described in Section 9.3.

### 9.2.2 Mathematical model of the heater

Electrical heaters are usually difficult to control as they require large power. There are basically two methods to control heaters. These are:

a. Phase angle control

b. Pulse width modulation control

*a. Phase angle control*

This is one of the common methods of power control where the start of each half-cycle is delayed by an angle. Thyristors (or triacs) are usually used in such circuits and the triggering angle is varied in order to change the power delivered to the heater element. Figure 9.2 shows a typical thyristor based phase angle control circuit. Assuming that the heater has a pure resistance R, and the supply voltage has a peak value $V_{max}$, it can be shown that the power delivered to the heater element is:

$$P = \frac{V_{rms}^2}{R} = \frac{V_{max}^2}{2\pi R}(\pi - \alpha + 0.5\sin 2\alpha) \tag{9-3}$$

where, each half cycle is delayed by an angle $\alpha$ as shown in Fig. 9.3.

*b. Pulse width modulation control*

In this method, the heater current is turned on an off as in a pulse width modulated waveform, where the period of the waveform is fixed but the ratio of the *on-time*

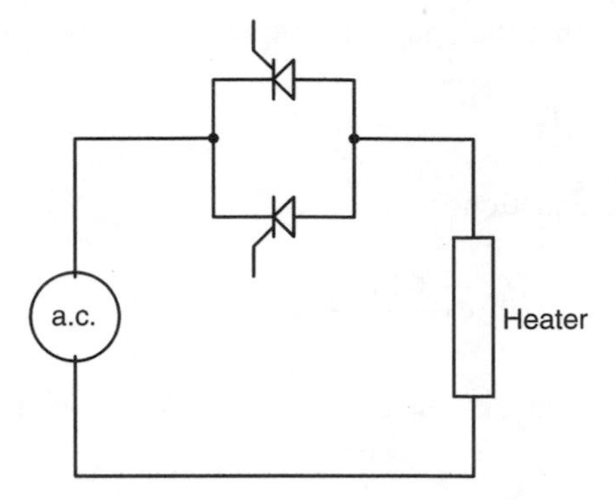


Fig. 9.2 Phase angle control of a heater

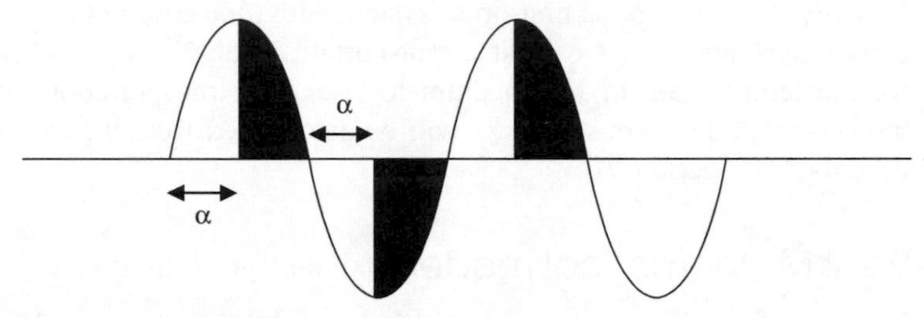


Fig. 9.3 Waveform of phase angle control

to *off-time* is varied according to the control voltage. If the applied voltage is a.c., the output is a burst of full-wave rectified pulses, where the pulse width of the on-time varies (see Fig. 9.4).

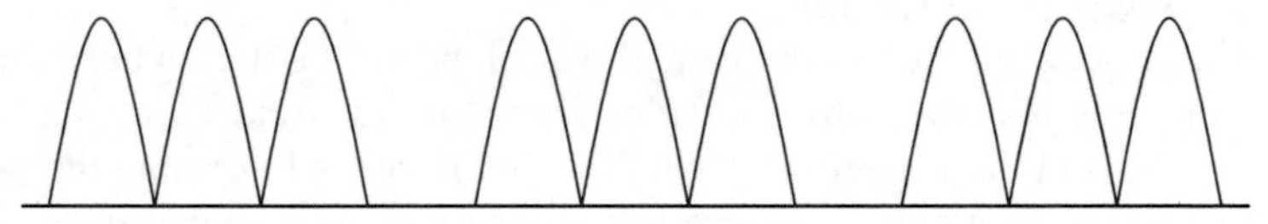

Fig. 9.4 Pulse width modulation of a.c.

In this project, a small 12 V heater is used. A pulse width modulated voltage is applied to the heater element in order to control the heating. The heater can be modelled as follows:

A pulse width modulated waveform is generated from the microcontroller as shown in Fig. 9.5, where M and S are the mark and the space of the waveform, and T is the period, i.e. $T = M + S$. This waveform is used to control a power MOSFET switch where the heater element is connected as the load of this device (see Fig. 9.6).

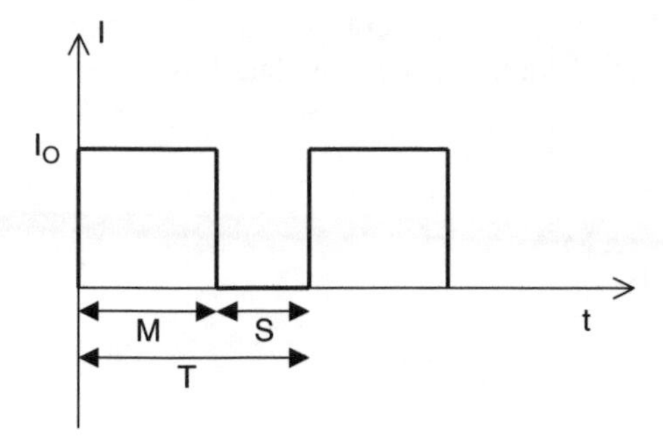


**Fig. 9.5** Mark (M) and space (S) of a PWM waveform

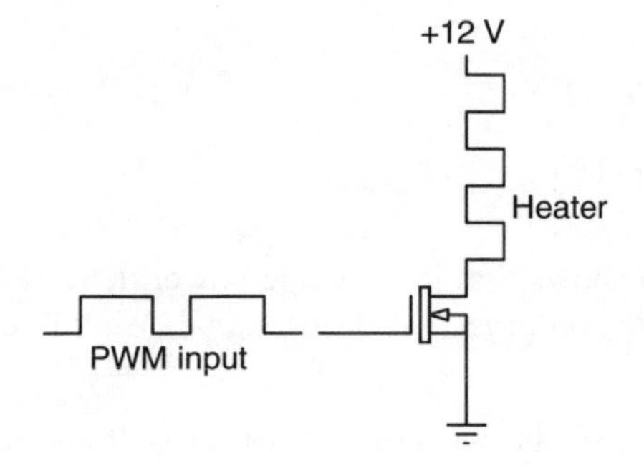


**Fig. 9.6** Circuit diagram of the heater driver

The rms value of the current through the heater element can be calculated as:

$$I_{rms} = \sqrt{\frac{1}{T}\int_0^T i^2(t)\,dt}$$

$$= \sqrt{\frac{1}{T}\int_0^M I_o^2}$$

$$= \sqrt{\frac{MI_o^2}{T}}$$

or,

$$I_{rms} = I_o\sqrt{\frac{M}{T}} \tag{9-4}$$

Assuming the heating element has a pure resistance, R, the average power delivered to the heater can be calculated as:

$$P_{AV} = RI_{rms}^2$$

$$= RI_o^2 \frac{M}{T}$$

if we let

$$\alpha = \frac{RI_o^2}{T}$$

then

$$P_{AV} = \alpha \cdot M \tag{9-5}$$

Equation 9-5 shows that the average power delivered to the load is linearly proportional to the *on-time* (M) of the signal. We will call M, the *duty cycle* of the waveform.

The frequency of the waveform must be well above the closed-loop bandwidth of the control system so that the process is affected by the mean level of the waveform. In this project, we will assume a frequency of 1 kHz, i.e. the period is 1 ms.

In this project,

$$R = 1.2\,\Omega$$

$$I_o = 10\,A$$

$$T = 1\,ms = 10^{-3}\,s$$

Thus, the transfer function of the heater is, from equation 9-5,

$$P_{AV} = \frac{1.2 \times 100}{10^{-3}} M$$

or,

$$\frac{P_{AV}}{M} = 1.2 \times 10^5 \tag{9-6}$$

where $P_{AV}$ is in Watts and M is in seconds.

### 9.2.3 Mathematical model of the temperature sensor

The temperature sensor used in the project is the LM35DZ (see Section 6.1.1). This sensor provides a 10 mV/°C analog output:

$$V_o = 10 \times 10^{-3} T$$

or,

$$V_o = 0.01\,T \qquad (9\text{-}7)$$

Where, $V_o$ is the sensor output voltage in Volts, and T is the temperature in °C.

## 9.3 The circuit diagram

The complete circuit diagram of the system is shown in Fig. 9.7. The temperature sensor is connected to analog channel AN0 of the microcontroller. Pulse width modulated output of the microcontroller (Pin CCP1) drives the gate input of a

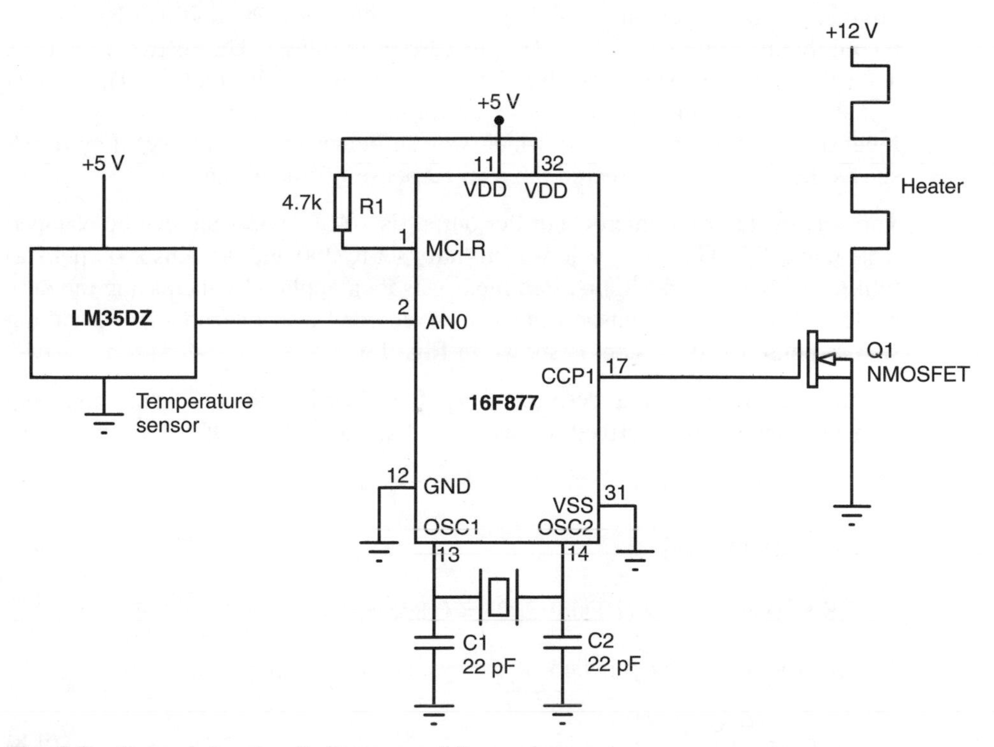


**Fig. 9.7** Complete circuit diagram of the system

power MOSFET transistor directly. The heater is connected to the drain input of the MOSFET. It is important to ensure that the MOSFET transistor chosen can dissipate the required maximum power. It may be necessary to mount the transistor on a heat-sink so that the required power can safely be dissipated without damaging the transistor. An IRL1004 type power MOSFET is used in this project. This transistor has a logic-level gate drive input and it can have a maximum continuous drain current of 130A. The maximum power which can be dissipated by this transistor is 200 W.

The desired temperature set point is hard-coded inside the control program.

## 9.4 Identification of the system

System identification was based on step input response of the open-loop system. A step input was applied to the heater driver by using the PWM output of the microcontroller. The temperature of the water in the tank was then measured and recorded every second by connecting the output of the LM35 sensor to the voltage input of a *DrDaq* hardware and the *Picolog* software. Both of these products are manufactured by PICO Technology. *DrDaq* is a small electronic card which is plugged into the parallel port of a PC. The card is equipped with sensors to measure the physical quantities such as the light intensity, sound level, voltage, humidity, and temperature. *Picolog* software runs on a PC and can be used to record the measurements of the *DrDaq* card in real-time. The software includes a graphical option which enables the measurements to be plotted. The results can also be saved in a spread-sheet type format for further analysis. The voltage input has a resolution of 5 mV which was sufficient for this project. Figure 9.8 shows the setup used to record the step response of the system.

The duty cycle of the microcontroller output is 10-bits wide and can be changed from 0 to 1023. The duty cycle was initially set to 200 and the sensor output stabilized at 265 mV (26.5°C). A step input was then applied by increasing the duty cycle to 1000 and the sensor voltage was recorded every second. The resulting step response of the system is shown in Fig. 9.9.

System parameters were derived using the Ziegler–Nichols open-loop step response method as described in Section 8.4.2. From Fig. 9.9:

$$T_d = 180\,\text{seconds}$$

$$T_1 = 1800\,\text{seconds}$$

$$K = (926 - 265)/(1000 - 200) = 0.826\,\text{mV}/\mu\text{s}$$

The open-loop transfer function of the system can be approximated to:

$$G(s) = \frac{0.826e^{-180\,s}}{(1 + 1800\,s)} \tag{9-8}$$

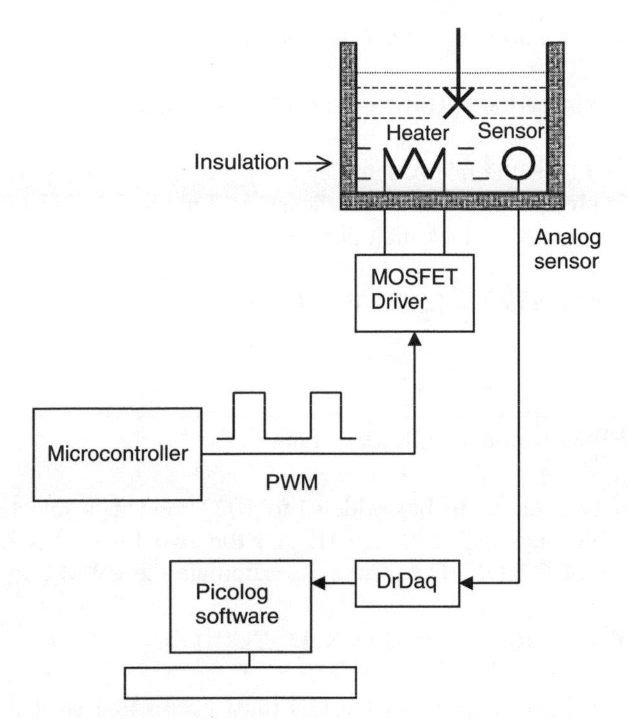


**Fig. 9.8** Setup used to record the step response of the system

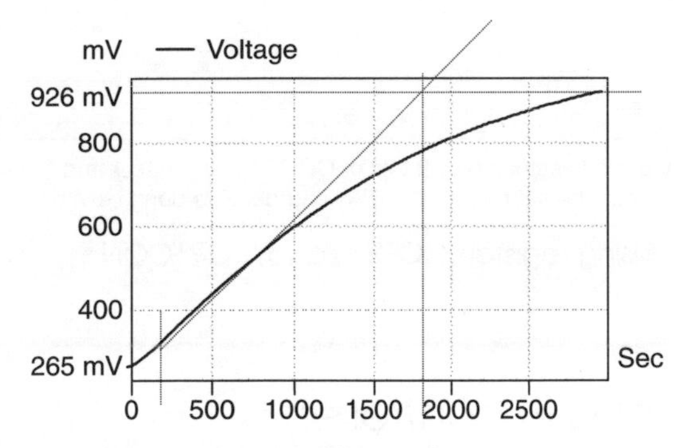


**Fig. 9.9** Open-loop step response of the system

# 9.5 Pulse width output of the microcontroller

PIC16F877 microcontroller provides two PWM outputs, known as CCP1 (Pin 17) and CCP2 (Pin 16). The PWM output CCP1 is controlled using Timer 2 and registers PR2, T2CON, CCPR1L, and CCP1CON.

The period of the PWM output CCP1 is set by loading a value into register PR2 and then selecting a clock multiplier value of either 1, 4, or 16. Equation 9-9 gives the formula for setting the period.

$$\text{PWM period} = (\text{PR2} + 1)^*4^*T^*_{osc}(\text{clock multiplier}) \quad (9\text{-}9)$$

Where $T_{osc}$ is the microcontroller clock period (0.250 μs with a 4 MHz crystal). In this project, the period was chosen as 1 ms by loading register PR2 with 249 and selecting the clock multiplier as 4, i.e.

$$\text{PWM period} = (249 + 1)^*4^*0.250^*4$$

or,

$$\text{PWM period} = 1000\,\mu s = 1\,ms$$

The duty cycle is 10 bits wide (0 to 1023) and it is selected by loading the 8 upper bits into register CCPR1L and the two lower bits into bits 4 and 5 of register CCP1CON. The formula to calculate the PWM duty cycle is:

$$\text{PWM duty cycle} = (\text{CCPR1L} : \text{CCP1CON})^*T^*_{osc}(\text{clock multiplier}) \quad (9\text{-}10)$$

Bits 2 and 3 of register CCP1CON must be set to 1 so that the microcontroller is in PWM mode. Figure 9.10 shows how the PWM duty cycle is selected by using CCPR1L and CCP1CON.

| CCPR1L | | | | | | | | CCP1CON | | | | | | | |
|---|---|---|---|---|---|---|---|---|---|---|---|---|---|---|---|
| 9 | 8 | 7 | 6 | 5 | 4 | 3 | 2 | X | X | 1 | 0 | 1 | 1 | X | X |

10-bit duty cycle is selected by 8 bits of CCPR1L and bits 4 and 5 of CCP1CON. Bits 2 and 3 of CCP1CON must be set to 1 for the PWM mode. X is don't care bit.

**Fig. 9.10** Using registers CCPR1L and CCP1CON

## 9.6 Design of a PI controller

Based on the Ziegler–Nichols open-loop test method, the parameters of the PI controller are given as:

$$K_p = \frac{0.9T_1}{KT_d} \qquad T_i = 3.3T_d$$

Thus, the PI parameters of our system are:

$$K_p = \frac{0.9 \times 1800}{0.826 \times 180} = 10.9 \qquad T_i = 3.3 \times 180 = 594$$

The transfer function of the required PI controller is:

$$\frac{U(s)}{E(s)} = 10.9\left[1 + \frac{1}{594\,s}\right]$$

or,

$$\frac{U(s)}{E(s)} = \frac{6474.6\,s + 10.9}{594\,s} \tag{9-11}$$

Figure 9.11 shows the block diagram of the closed-loop system with PI controller.

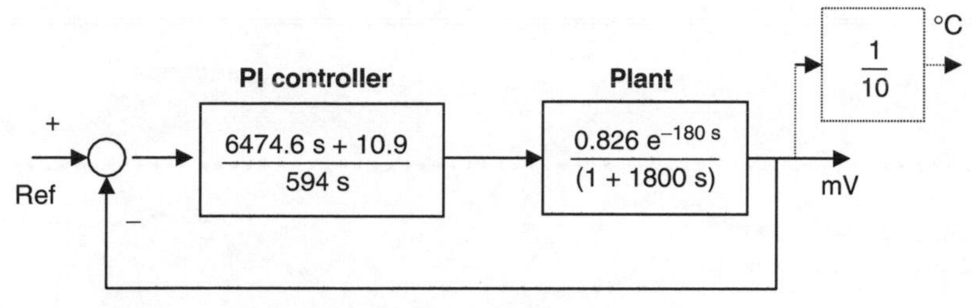


**Fig. 9.11** Closed-loop system with PI controller

## 9.6.1 The software of the PI controller

Using equation 8-13, the following PI parameters are used in the realization of the PI controller:

$$a = K_p \qquad b = \frac{K_p T}{T_i}$$

The system dominant time constant is 1800 seconds and it is recommended that the sampling time should be much less than 1800/10 = 180 seconds. Two different sampling times were chosen: 20 seconds and 60 seconds. The PI parameters for each sampling time are:

For T = 20 seconds:

$$a = 10.9 \qquad b = \frac{10.9 \times 20}{594} = 0.37$$

For T = 60 seconds:

$$a = 10.9 \qquad b = \frac{10.9 \times 60}{594} = 1.10$$

The set point was chosen as 300 mV, i.e. 30°C.

The complete program listing of the PI controller is given in Fig. 9.12. *MIN* and *MAX* are the minimum and the maximum values of the controller output and are set to 0 and 1000 respectively. The PI parameters a and b are then

```
/*********************************************************************
*
*
*     PROJECT:     PI
*     FILE:        PI.C
*     DATE:        November 2001
*     PROCESSOR:   PIC16F877
*     COMPILER:    FED C
*
*
* This program implements the PI algorithm for the project given
* in Chapter 9.
*
*
* The PI parameters are:      a = 10.9
*                                        b = 0.37
*                                        T = 20 seconds
*                                        R = 300 mV (30C)
*
*********************************************************************/

#include <P16F877.h>
#include <delays.h>

void main()
{
            float a,b,set_point,rkt,LSB,ekt,pkt,qkt,ykt,ukt;
            float MAX,MIN,pkt_1,ekt_1;
            int control;

            LSB=5000.0/1024.0;
            MIN=0.0;
            MAX=1000.0;
            pkt_1=0.0;
            ekt_1=0.0;

            /* Define PID parameters */
            a=10.9;
            b=0.37;

            /* Required set-point temperature (in mV. 10mV/C) */
            set_point=300.0;

            /* Configure Port C as output for the PWM */
            TRISC=0;

            /* Define PWM period and enable PWM mode */
            PR2=249;                                  //PWM period=1ms
            CCP1CON=0x3C;                             //enable PWM mode
            T2CON=5;                                  //Timer2 on with
multiplier=4

            /* START OF PROPORTIONAL-INTEGRAL LOOP
```

**Fig. 9.12** Program listing of the PI controller

```
   ================================== */
    while(1)
    {
          /* Configure the A/D */
          ADCON1=0x80;                              //6 MSB bits to zero
          ADCON0=0x41;                              //set A/D oscillator

          /* Start A/D conversion */
          ADCON0=0x45;
          while((ADCON0 & 4) != 0);      //wait for conversion

          /* Now read in the A/D output */
          ykt=256.0*(float)ADRESH+(float)ADRESL;
          ykt=ykt*LSB;                              //sensor output in mV

          rkt=set_point;

          /* Calculate error */
          ekt=rkt-ykt;

          /* Calculate I term */
          pkt=b*ekt+pkt_1;


          /* Calculate PI output */
          ukt=pkt+a*ekt;

          if(ukt > MAX)
          {
                pkt=pkt_1;
                ukt=MAX;
          }
          else if(ukt < MIN)
          {
                pkt=pkt_1;
                ukt=MIN;
          }

          /* Send control to heater driver */
          ukt=ukt-3.0;
          ukt=ukt/4.0;
          control=ukt;
          CCPR1L=control;

          /* Save variables */
          pkt_1=pkt;
          ekt_1=ekt;

          /* Wait for 20 second */
          Wait(20000);
    }

}
```

**Fig. 9.12** (*Continued*)

defined and the set point is chosen as 300 mV. The PWM mode is enabled and Timer2 is set to run with a clock multiplier of 4. The A/D converter is then started and the output of the temperature sensor is read and converted into mV. The error term *ekt* is calculated from the difference of the set point *rkt* and the output *ykt*. The integral action *pkt* is calculated and this is added to the proportional term *a*ekt* in order to obtain the controller output *ukt*. If this output is greater than the maximum allowable value *MAX*, the output is set to *MAX*, and the integral action is stopped to avoid the integral wind-up. If on the other hand the controller output is less than *MIN*, the output is set to *MIN* and again the integral action is stopped. The controller output *ukt* is then loaded into the PWM register CCPR1L. Note that *ukt* is divided by 4 since CCPR1L is the higher 8-bits of the 10-bit PWM value. The lower two bits are set to 1 in register CCP1CON at the beginning of the program. The control loop is repeated after a delay of 20 seconds. Note that the actual sampling time here is slightly higher than 20 seconds because of the time taken to process the PI loop. An exact 20 seconds sampling time can be obtained by using software timer interrupt techniques and including the PI algorithm as part of this interrupt service routine.

#### 9.6.1.1 Results

Figure 9.13 shows the system response when T = 20 seconds. The response with T = 60 seconds is shown in Fig. 9.14. It is clear that the two responses are very similar and the temperature is settled to 30°C as expected.

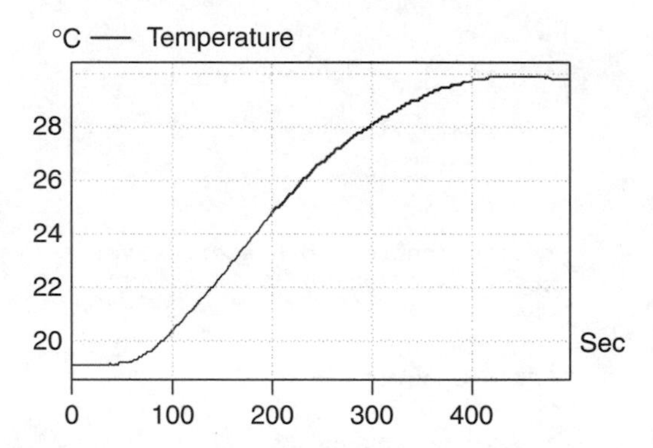


**Fig. 9.13** System response when T = 20 seconds

A trial output was obtained by doubling the integral action time, i.e. a = 10.9, b = 0.74, and T = 20 seconds. This response is shown in Fig. 9.15 and it is clear here that the rise time is faster, but the response has an overshoot and a longer settling time.

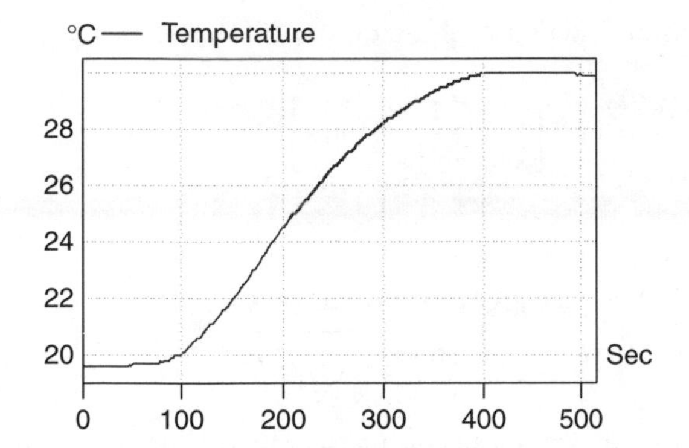


**Fig. 9.14** System response when T = 60 seconds

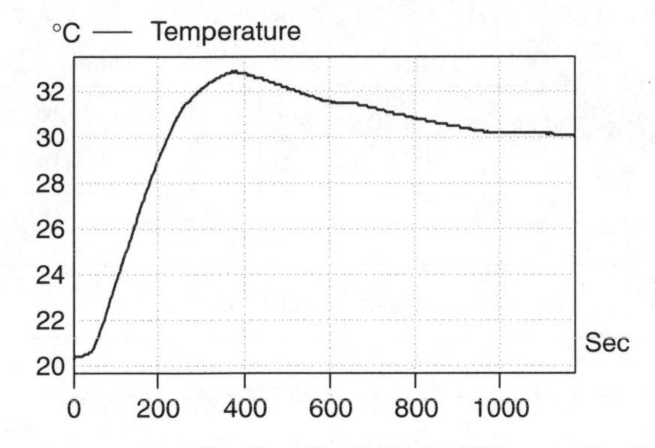


**Fig. 9.15** System response with doubled integral time, when T = 20 seconds

# 9.7 Design of a PID controller

Based on the Ziegler–Nichols open-loop test method, the parameters of the PID controller are given as:

$$K_p = \frac{1.2T_1}{KT_d} \qquad T_i = 2T_d \qquad T_D = 0.5T_d$$

Thus, from equation 9-8,

$$K_p = \frac{1.2 \times 1800}{0.826 \times 180} = 14.5 \qquad T_i = 2 \times 180 = 360$$

$$T_D = 0.5 \times 180 = 90$$

The transfer function of the required PID controller is:

$$\frac{U(s)}{E(s)} = 14.5\left[1 + \frac{1}{360\,s} + 90\,s\right]$$

or,

$$\frac{U(s)}{E(s)} = \frac{469\,800\,s^2 + 5220\,s + 14.5}{360\,s} \tag{9-12}$$

Figure 9.16 shows the block diagram of the closed-loop system.

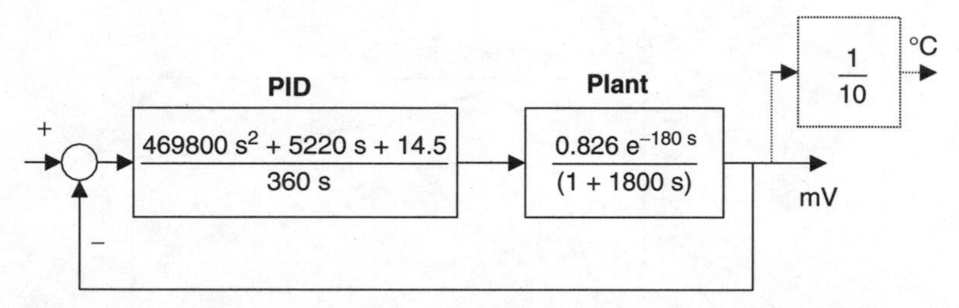


**Fig. 9.16** Closed-loop system with PID controller

## 9.7.1 The software of the PID controller

Using equation 8-13, the parameters of the PID controller are:

$$a = K_p \qquad b = \frac{K_p T}{T_i} \qquad c = \frac{K_p T_D}{T}$$

or, if we let $T = 60$ seconds, then,

$$a = 14.5 \qquad b = \frac{14.5 \times 60}{360} = 2.4 \qquad c = \frac{14.5 \times 180}{60} = 43.5$$

The PID algorithm given in Section 8.5 is used and the complete program listing is shown in Fig. 9.17. The program is very similar to the PI controller program with the exception that here the derivative action (*qkt*) is added to the program.

### 9.7.1.1 Results

Figure 9.18 shows the system response, and as expected, the temperature is settled to 30°C. In this example, the PID response is not as good as the PI response.

```
/***********************************************************************
*
*
*       PROJECT:     PID
*       FILE:        PID.C
*       DATE:        November 2001
*       PROCESSOR:   PIC16F877
*       COMPILER:    FED C
*
*
* This program implements the PID algorithm for the project given
* in Chapter 9.
*
*       The PID parameters are:
*                          a = 14.5
*                          b = 2.4
*                          c = 43.5
*                          T = 60 seconds
*                          R = 300 mV (30C)
*
***********************************************************************/

#include <P16F877.h>
#include <delays.h>


void main()
{
            float a,b,c,set_point,rkt,LSB,ekt,pkt,qkt,ykt,ukt;
            float MAX,MIN,pkt_1,ekt_1;
            int control,i;

            LSB=5000.0/1024.0;
            MIN=0.0;
            MAX=1000.0;
            pkt_1=0.0;
            ekt_1=0.0;

            /* Define PID parameters */
            a=14.5;
            b=2.4;
            c=43.5;

            /* Required set-point temperature (in degrees Celsius) */
            set_point=300.0;

            /* Configure Port C as output for the PWM */
            TRISC=0;

            /* Define PWM period and enable PWM mode */
            PR2=249;                                  //PWM period=1ms
            CCP1CON=0x3C;                             //enable PWM mode
            T2CON=5;                                  //Timer2 on with
      multiplier=4
```

Fig. 9.17 Program listing of the PID controller

```
/* START OF PROPORTIONAL-INTEGRAL-DERIVATIVE LOOP
   ============================================== */
while(1)
{
     /* Configure the A/D */
     ADCON1=0x80;                                  //6 MSB bits to zero
     ADCON0=0x41;                                  //set A/D oscillator

     /* Start A/D conversion */
     ADCON0=0x45;
     while((ADCON0 & 4) != 0);      //wait for conversion

     /* Read the sensor output */
     ykt=256.0*(float)ADRESH+(float)ADRESL;
     ykt=ykt*LSB;                                  //sensor output in mV

     rkt=set_point;

     /* Calculate error */
     ekt=rkt-ykt;

     /* Calculate I term */
     pkt=b*ekt+pkt_1;

     /* Calculate D term */
     qkt=c*(ekt-ekt_1);

     /* Calculate PID output */
     ukt=pkt+a*ekt+qkt;

     if(ukt > MAX)
     {
          pkt=pkt_1;
          ukt=MAX;
     }
     else if(ukt < MIN)
     {
          pkt=pkt_1;
          ukt=MIN;
     }

     /* Send control to heater driver */

     ukt=ukt-3.0;
     ukt=ukt/4.0;
     control=ukt;
     CCPR1L=control;

     /* Save variables */
     pkt_1=pkt;
     ekt_1=ekt;

     /* Wait for 60 seconds */
     Wait(60000);
}
}
```

**Fig. 9.17** (*Continued*)

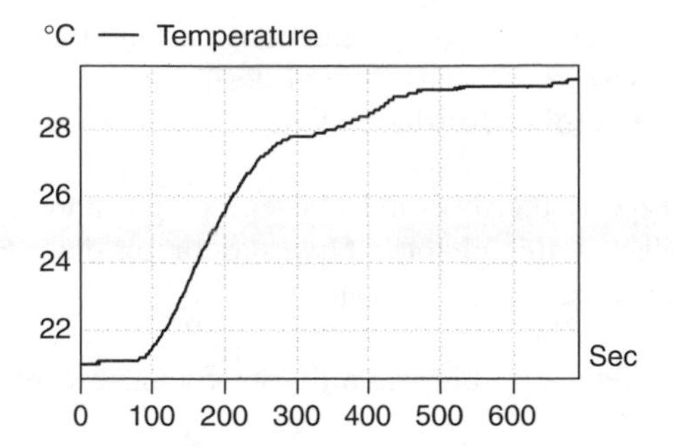


**Fig. 9.18** System response with PID controller

## 9.8 Compensating for heat losses

The PID (or PI) controller suggests that in the steady state, when the error is zero, the controller output is also zero, which provides no voltage to the heater driver. However, to compensate for the heat losses from the water tank, an input is needed to the heater. The integral term of the controller normally compensates for this loss, but there is usually difficulties in maintaining a smooth output. The normal practice is to modify the PID (or the PI) algorithm by adding a constant term $M$ to the right-hand side of the equation. $M$ is independent of the error signal and it represents the value of the manipulated variable at the steady state operating point. The value of $M$ is normally calculated for a given steady state operating point and is inserted as a constant in the program.

## 9.9 Other considerations

What is given in this chapter is the design of a small temperature control system. In industrial applications there are usually large temperature control systems with multiple sensors and high-power heating elements. A.C. power control techniques (based on thyristor and triac circuits) are usually employed to control the power delivered to the load. Design of industrial temperature control systems require the analysis of both the heating and the cooling behaviour of the plant. The control of such systems is usually a complex task and requires considerable experience both in electronic engineering and in control engineering. Safety and reliable operation of such plants are also very important factors which must be considered during the design phase.

## 9.10 Exercises

1. Design a bang-bang type controller for the temperature control system described in this chapter. Give the complete program listing of the controller.

2. Derive the continuous system transfer function of the system given in this chapter. Use continuous system design techniques to design a suitable continuous controller for the system.

3. Use a type K thermocouple instead of the integrated circuit sensor used in the project in this chapter. Give the circuit diagram of the complete system and the program listing, assuming a PIC16F877 type microcontroller is used.

4. Repeat exercise 3, but use a thermistor instead of a thermocouple.

5. Repeat exercise 3, but use a RTD sensor instead of a thermocouple.

# Appendix A

# Platinum RTD Sensor Resistances ($\alpha = 0.00385$)

| | 0 | 1 | 2 | 3 | 4 | 5 | 6 | 7 | 8 | 9 | 10 |
|---|---|---|---|---|---|---|---|---|---|---|---|
| **0** | 100.00 | 100.39 | 100.78 | 101.17 | 101.56 | 101.95 | 102.34 | 102.73 | 103.12 | 103.51 | 103.90 |
| **10** | 103.90 | 104.29 | 104.68 | 105.07 | 105.46 | 105.85 | 106.24 | 106.63 | 107.02 | 107.40 | 107.79 |
| **20** | 107.79 | 108.18 | 108.57 | 108.96 | 109.35 | 109.73 | 110.12 | 110.51 | 110.90 | 111.29 | 111.67 |
| **30** | 111.67 | 112.06 | 112.45 | 112.83 | 113.22 | 113.61 | 114.00 | 113.38 | 114.77 | 115.15 | 115.54 |
| **40** | 115.54 | 115.93 | 116.31 | 116.70 | 117.08 | 117.47 | 117.86 | 118.24 | 118.63 | 119.01 | 119.40 |
| **50** | 119.40 | 119.78 | 120.17 | 120.55 | 120.94 | 121.32 | 121.71 | 122.09 | 122.47 | 122.86 | 123.24 |
| **60** | 123.24 | 123.63 | 124.01 | 124.39 | 124.78 | 125.16 | 125.54 | 125.93 | 126.31 | 126.69 | 127.08 |
| **70** | 127.08 | 127.46 | 127.84 | 128.22 | 128.61 | 128.99 | 129.37 | 129.75 | 130.13 | 130.52 | 130.90 |
| **80** | 130.90 | 131.28 | 131.66 | 132.04 | 132.42 | 132.80 | 133.18 | 133.57 | 133.95 | 134.33 | 134.71 |
| **90** | 134.71 | 135.09 | 135.47 | 135.85 | 136.23 | 136.61 | 136.99 | 137.37 | 137.75 | 138.13 | 138.51 |

# Appendix B

# ASCII Code

ASCII codes of the first 128 characters are standard and the same code is used between different equipment manufacturers. ASCII codes of characters between 128 and 255 are also known as the extended ASCII characters and these characters and their codes may differ between different computer manufacturers. Below is a list of the most commonly used ASCII characters and their codes both in hexadecimal and in binary.

| Character | Hex | Binary | Character | Hex | Binary |
|---|---|---|---|---|---|
| NUL | 00 | 00000000 | DLE | 10 | 00010000 |
| SOH | 01 | 00010001 | XON | 11 | 00010001 |
| STX | 02 | 00100010 | DC2 | 12 | 00010010 |
| ETX | 03 | 00110011 | XOFF | 13 | 00010011 |
| EOT | 04 | 01000100 | DC4 | 14 | 00010100 |
| ENQ | 05 | 10011001 | NAK | 15 | 00011001 |
| ACK | 06 | 01100110 | SYN | 16 | 00010110 |
| BEL | 07 | 01110111 | ETB | 17 | 00010111 |
| BS | 08 | 10001000 | CAN | 18 | 00011000 |
| HT | 09 | 10011001 | EM | 19 | 00011001 |
| LF | 0A | 10101010 | SUB | 1A | 00011010 |
| VT | 0B | 10111011 | ESC | 1B | 00011011 |
| FF | 0C | 11001100 | FS | 1C | 00011100 |
| CR | 0D | 11011101 | GS | 1D | 00011101 |
| SO | 0E | 11101110 | RS | 1E | 00011110 |
| SI | 0F | 11111111 | US | 1F | 00011111 |

| Character | Hex | Binary | Character | Hex | Binary |
|---|---|---|---|---|---|
| SP | 20 | 00100000 | ; | 3B | 00111011 |
| ! | 21 | 00100001 | < | 3C | 00111100 |
| " | 22 | 00100010 | = | 3D | 00111101 |
| # | 23 | 00100011 | > | 3E | 00111110 |
| $ | 24 | 00100100 | ? | 3F | 00111111 |
| % | 25 | 00101001 | @ | 40 | 01000000 |
| & | 26 | 00100110 | A | 41 | 01000001 |
| ' | 27 | 00100111 | B | 42 | 01000010 |
| ( | 28 | 00101000 | C | 43 | 01000011 |
| ) | 29 | 00101001 | D | 44 | 01000100 |
| * | 2A | 00101010 | E | 45 | 01001001 |
| + | 2B | 00101011 | F | 46 | 01000110 |
| , | 2C | 00101100 | G | 47 | 01000111 |
| – | 2D | 00101101 | H | 48 | 01001000 |
| . | 2E | 00101110 | I | 49 | 01001001 |
| / | 2F | 00101111 | J | 4A | 01001010 |
| 0 | 30 | 00110000 | K | 4B | 01001011 |
| 1 | 31 | 00110001 | L | 4C | 01001100 |
| 2 | 32 | 00110010 | M | 4D | 01001101 |
| 3 | 33 | 00110011 | N | 4E | 01001110 |
| 4 | 34 | 00110100 | O | 4F | 01001111 |
| 5 | 35 | 00111001 | P | 50 | 10010000 |
| 6 | 36 | 00110110 | Q | 51 | 10010001 |
| 7 | 37 | 00110111 | R | 52 | 10010010 |
| 8 | 38 | 00111000 | S | 53 | 10010011 |
| 9 | 39 | 00111001 | T | 54 | 10010100 |
| : | 3A | 00111010 | U | 55 | 10011001 |

<table>
<tr><th>Character</th><th>Hex</th><th>Binary</th><th>Character</th><th>Hex</th><th>Binary</th></tr>
<tr><td>V</td><td>56</td><td>10010110</td><td>k</td><td>6B</td><td>01101011</td></tr>
<tr><td>W</td><td>57</td><td>10010111</td><td>l</td><td>6C</td><td>01101100</td></tr>
<tr><td>X</td><td>58</td><td>10011000</td><td>m</td><td>6D</td><td>01101101</td></tr>
<tr><td>Y</td><td>59</td><td>10011001</td><td>n</td><td>6E</td><td>01101110</td></tr>
<tr><td>Z</td><td>5A</td><td>10011010</td><td>o</td><td>6F</td><td>01101111</td></tr>
<tr><td>[</td><td>5B</td><td>10011011</td><td>p</td><td>70</td><td>01110000</td></tr>
<tr><td>\</td><td>5C</td><td>10011100</td><td>q</td><td>71</td><td>01110001</td></tr>
<tr><td>]</td><td>5D</td><td>10011101</td><td>r</td><td>72</td><td>01110010</td></tr>
<tr><td>^</td><td>5E</td><td>10011110</td><td>s</td><td>73</td><td>01110011</td></tr>
<tr><td>_</td><td>5F</td><td>10011111</td><td>t</td><td>74</td><td>01110100</td></tr>
<tr><td>`</td><td>60</td><td>01100000</td><td>u</td><td>75</td><td>01111001</td></tr>
<tr><td>a</td><td>61</td><td>01100001</td><td>v</td><td>76</td><td>01110110</td></tr>
<tr><td>b</td><td>62</td><td>01100010</td><td>w</td><td>77</td><td>01110111</td></tr>
<tr><td>c</td><td>63</td><td>01100011</td><td>x</td><td>78</td><td>01111000</td></tr>
<tr><td>d</td><td>64</td><td>01100100</td><td>y</td><td>79</td><td>01111001</td></tr>
<tr><td>e</td><td>65</td><td>01101001</td><td>z</td><td>7A</td><td>01111010</td></tr>
<tr><td>f</td><td>66</td><td>01100110</td><td>{</td><td>7B</td><td>01111011</td></tr>
<tr><td>g</td><td>67</td><td>01100111</td><td>|</td><td>7C</td><td>01111100</td></tr>
<tr><td>h</td><td>68</td><td>01101000</td><td>}</td><td>7D</td><td>01111101</td></tr>
<tr><td>i</td><td>69</td><td>01101001</td><td>~</td><td>7E</td><td>01111110</td></tr>
<tr><td>j</td><td>6A</td><td>01101010</td><td>(delete)</td><td>7F</td><td>01111111</td></tr>
</table>

# Appendix C

# FED C Compiler Library Functions

A list of the FED C Compiler library functions are given in this appendix. More detailed information on these functions can be obtained from Forest Electronic Developments (web site: http://www.fored.co.uk and e-mail: info@fored.co.uk).

1. ClockDataIn

This function is used to clock in serial data from any pin of the PIC microcontroller. The port name, character count, clock port number, and the data port numbers are specified by the user.

2. ClockDataOut

This function is used to clock out serial data from any pin of the PIC microcontroller. The port name, character count, clock port number, and the data port numbers are specified by the user.

3. 1 Wire Bus control

These functions are used to provide interface to the *Dallas 1 Wire Bus*. Functions are provided to reset the bus, check for devices on the bus, and transmit and receive data from other devices on the bus.

4. EEPROM memory control

There are functions to read and write a byte to the EEPROM memory of a PIC microcontroller which is equipped with EEPROM type data memory. Function *ReadEEData* reads a data byte and function *WriteEEData* writes a byte to the EEPROM data memory.

5. Infra red functions

These functions enable a byte to be sent and received from any pin of the PIC microcontroller in IRDA format. Function *IRTx* sends out a byte and function *IRRx* receives a byte.

6. $I^2C$ bus functions

These functions read and write a byte to the $I^2C$ bus. Function *IIRead* reads a byte from the $I^2C$ bus and function *IIWrite* writes a byte to the $I^2C$ bus.

7. Interrupt driven serial I/O

These functions are used to drive the serial hardware of the PIC microcontrollers which are equipped with USART type serial I/O hardware (e.g. PIC16F877). Function *AddTx* adds a byte to the transmit buffer. Function *GetRxSize* returns the number of bytes in the receive buffer. Function *GetTxSize* returns the number of bytes in the transmit buffer. Function *SerIntInit* initializes the serial port.

8. Hex keypad functions

These functions scan a 4 × 4 keypad where the rows are outputs from a PIC microcontroller and columns are inputs to the microcontroller. The function returns a key number from 0 to 15 if a key is pressed.

9. LCD functions

Several functions are provided to control a parallel LCD and also send data to a LCD. Function *LCD* can be used to control the position of the cursor and to clear the LCD screen. Function *LCDString* can be used to send a string of data to the LCD display.

10. Maths functions

A number of floating point maths functions are provided which can be used to perform trigonometric and other mathematical calculations. The following maths functions are provided:

- cos — cosine
- sin — sine
- tan — tangent
- sqrt — square root
- exp — returns e to the power y
- exponent — returns the exponent of a floating point number
- log — natural logarithm
- log10 — logarithm to base 10
- mantissa — returns the mantissa of a floating point number
- pow — returns the power of a number
- e — macro (2.71828...)
- PI — PI (3.14159...)
- PowerSeries — calculates the sum of a power series
- fabs — absolute value of a variable
- fPrtString — prints a floating point number in string format

11. printf functions

Several functions are provided to print numbers, characters and strings in user formats.

12. Random number generator

Functions are provided to generate pseudo random numbers.

13. Serial IN/OUT

Function *SerialIn* is used to receive asynchronous serial data using any pin of a PIC microcontroller. Similarly, function *SerialOut* is used to send out a data byte in serial asynchronous format. The data bit rate must be defined before a data can be received or sent.

14. String functions

A large number of string functions are given for manipulating string variables. A list of the string functions are given below:

- CheckSum returns the sum of all the characters
- memcpy copy a number of bytes from one location to another location
- strcat add two strings together
- strchr find a character in a string
- strcmp compare two strings
- strcpy copy one string into another string
- vstrcpy copy one string into another string
- strlen returns the length of a string
- strlwr convert characters in a string to lower case
- strupr convert characters in a string to upper case
- rCheckSum sum all characters in a string in RAM
- rmemcpy Copy a number of bytes from one location to another location in RAM
- rstrcat add one string to the end of another one in RAM
- rstrchr find a character in a string in RAM

15. String print functions

A number of functions are provided to print a decimal representation of a signed number to a string. For example, an integer can be printed into a string format using the function *iPrtString*.

# Glossary

**ADC** Analogue-to-digital converter. A device that converts analogue signals to a digital form for use by a computer.

**Algorithm** A fixed step-by-step procedure for finding a solution to a problem.

**Alumel** An aluminum nickel alloy used in the negative leg of a type K thermocouple.

**Ambient temperature** The temperature of the surrounding air which is in contact with the equipment or the devices under test.

**ANSI** American National Standards Institute.

**Anti-alias filter** An anti-alias filter allows through the lower frequency components of a signal but stops higher frequencies. Anti-alias filters are specified according to the sampling rate of the system.

**ASCII** American Standard Code for Information Interchange. A widely used code in which alphanumeric characters and certain other special characters are represented by unique 7-bit binary numbers. For example, the ASCII code of letter "A" is 65.

**Bidirectional port** An interface port that can be used to transfer data in either direction.

**Binary** The representation of numbers in base two system.

**Bit** A single binary digit.

**Byte** A group of 8 binary digits.

**Callendar–van Duesen equation** An equation which provides resistance values as a function of temperature for RTDs.

**Celsius** A temperature scale defined by 0° at the ice point and 100° at the boiling point of water.

**Chromel** A chromium-nickel alloy which is used in the positive leg of types K and E thermocouples.

**Clock** A circuit generating regular timing signals for a digital logic system. In a microcomputer system clocks are usually generated using crystal devices. A typical clock frequency is 12 MHz.

**CMOS** Complementary Metal Oxide Semiconductor. A family of integrated circuits that offers extremely high packing density and low power.

**Cold junction** The reference junction of a thermocouple which is kept at a constant temperature.

**Cold junction compensation** Compensation for thermocouple reference junction temperature variations.

**Compiler** A program designed to translate high-level languages into machine code.

**Constantan** A copper-nickel alloy used as the negative leg in Types E, J, and T thermocouples.

**Cycle time** Time required to access a memory location or to carry out an operation in a computer system.

**DAC** Digital to Analogue Converter. A device that converts digital signals into analogue form.

**Execute** To perform a specified operation sequence in a program.

**Fahrenheit** A temperature scale defined by 32° at the ice point and 212° at the boiling point of water at sea level.

**Flowchart** Graphical representation of the operation of a program.

**Hardware** The physical parts or electronic circuitry of a computer system.

**Hexadecimal** Base 16 numbering system. In hexadecimal notation, numbers are represented by digits 0–9 and the characters A–F. For example, decimal number 165 is represented as A5.

**High-level language** Programming language in which each instruction or statement corresponds to several machine code instructions. Some high-level languages are BASIC, FORTRAN, C, PASCAL and so on.

**Input port** Part of a computer that passes external signals into a computer. Microcomputer input ports are usually 8-bits wide.

**I/O** Short for Input/Output.

**Interface** To interconnect a computer to external devices and circuits.

**Interrupt** An external or internal event that suspends the normal program flow within a computer and causes entry into a special interrupt program (also called the interrupt service routine). For example, an external interrupt could be generated when a button is pressed. An internal interrupt could be generated when a timer reaches a certain value.

**ISR** Interrupt Service Routine. A program that is entered when an external or an internal interrupt occurs. Interrupt service routines are usually high-priority routines.

**Junction** The point in a thermocouple where the two dissimilar metals are joined.

**Kelvin** Absolute temperature scale based on the Celsius scale, but zero K is defined at absolute zero and 0°C corresponds to 273.15°K.

**LCD** Liquid Crystal Display. A low-powered display which operates on the principle of reflecting incident light. An LCD does not itself emit light. There are many varieties of LCDs. For example, numeric, alphanumeric, or graphical.

**LED** Light Emitting Diode. A semiconductor device that emits a light when a current is passed in the forward direction. There are many colours of LEDs. For example, red, yellow, green, and white.

**Microcomputer** General purpose computer using a microprocessor as the CPU. A microcomputer consists of a microprocessor, memory, and input/output.

**Microprocessor** A single large-scale integrated circuit which performs the functions of a CPU.

**Multiplexing** The technique where each signal is switched in turn to a single analogue-to-digital converter. As opposed to where one converter is used for each signal in multi-sampling.

**Negative temperature coefficient** A decrease in resistance with an increase in temperature.

**On/off controller** A controller whose action is fully on or fully off.

**Output port** Part of a computer which is used to pass electrical signals to outside the computer. Microcomputer output ports are usually 8 bits wide.

**PDL** Program Description Language. Representation of the control and data flow in a program using simple English-like sentences.

**PID** Proportional, Integral, and Derivative control algorithm.

**Platinum** The negative leg in Types R and S thermocouples.

**Positive temperature coefficient** An increase in resistance with an increase in temperature.

**Pull down resistor** A resistor connected between an I/O pin and ground.

**Pull up resistor** A resistor connected to the output of an open collector (or open drain) transistor of a gate in order to load the output.

**Reference junction** The cold junction in a thermocouple circuit which is held at a known temperature. The standard reference temperature is 32°F.

**RTD** Abbreviation for Resistance Temperature Detector. It is a circuit element whose resistance increases with increasing temperature.

**Seeback coefficient** The rate of change of emf with temperature in a thermocouple junction.

**Self heating** Heating effect due to current flow in the sensing element of a resistance thermometer.

**Sensor** A device that can detect a change in a physical quantity and produce a corresponding electrical signal.

**Set point** The temperature at which a controller is set to control a system.

**Thermistor** A temperature-sensing element which exhibits a large change in resistance proportional to a small change in temperature.

**Thermocouple** A temperature sensor formed by joining two dissimilar metals and applying a temperature differential between the measuring junction and the reference point.

**Transducer** A device that converts a measurable quantity into an electronic signal. For example, a temperature transducer gives out an electrical signal which may be proportional to the temperature.

**Triple point** The temperature at which all three phases of a substance are in equilibrium. The triple point of water is 0.01°C.

**Wheatstone bridge** An electrical bridge circuit which can be used to measure resistance of a thermal sensor.

**Word** A group of 16 binary digits.

**Zero power resistance** The resistance of a thermistor or RTD element with no power being dissipated.

# Index